THE
ACCIDENTAL
SPECIES

The Accidental Species

MISUNDERSTANDINGS OF HUMAN EVOLUTION

Henry Gee

The University of Chicago Press

Chicago and London

The University of Chicago Press, Chicago 60637
The University of Chicago Press, Ltd., London

Paperback edition 2015
Printed in the United States of America

24 23 22 21 20 19 18 17 16 15   4 5 6 7 8

ISBN-13: 978-0-226-28488-0 (cloth)
ISBN-13: 978-0-226-27120-0 (paper)
ISBN-13: 978-0-226-04498-9 (e-book)
10.7208/chicago/9780226044989.001.0001

Library of Congress Cataloging-in-Publication Data

Gee, Henry, 1962– author.
The accidental species : misunderstandings of human evolution / Henry Gee
pages cm
Includes bibliographical references and index.
ISBN 978-0-226-28488-0 (cloth : alkaline paper)
ISBN 978-0-226-04498-9 (e-book)
1. Human evolution. 2. Human beings. I. Title.
GN281.G36 2013
599.93'8—dc23
2013016599

♾ This paper meets the requirements of ANSI/NISO Z39.48-1992 (Permanence of Paper)

To the memory of John Maddox (1925–2009):
colleague, mentor, and friend,
in the hope that he'd have approved

Contents

Preface: No More Missing Links

Here's the thing. It's the curious phenomenon in which otherwise sane and rational news reporters lose all sense of reason or proportion when confronted with anything to do with human evolution, no matter how trivial or (ultimately) inconsequential it might be. Scientists make all kinds of discoveries every day, but almost all add just one small brick to a wall of knowledge that's sky high. Very few are deserving of any press coverage at all, let alone in the tumescent tones reserved for human evolution. Yet it seems that any paper on human evolution is fair game for the breathlessly orgasmic treatment usually reserved for voice-overs for commercials for expensive ice cream. If all discoveries are treated the same way, one is forced to wonder, then no discrimination can be made between them, and the effect is a kind of dull infantilization in which the significance of the discovery is obscured, and science as a whole is done a disservice.

A recent case was the media brouhaha surrounding the discovery of a fossil primate called *Darwinius masillae*. If you care to look up the scientific paper in which *Darwinius* is described, you can—it's freely accessible to anyone.[1] If you do, you'll find a perfectly fine description of a rare and beautiful fossil. If you read carefully, you'll see that *Darwinius masillae* is one of a number of primates belonging to an extinct group called adapids. *Darwinius* is a particularly fine specimen of an adapid, but it does not reveal any exceptional insight into the evolution of adapids or of primates as a whole. The evolutionary significance of adapids is debated by specialists, but most agree that they are more closely related to lemurs and bush babies than modern monkeys or apes, let alone humans.

The media circus (there is no other word) implied something rather different—that the fossil represented a crucial stage in human evolution. It was the "link." It's a fair bet that whereas most people won't have read the scientific paper, with all its technical terminology, they might very well have seen the TV special and accompanying book, launched in a blaze of flashbulbs.

It is partly because of this that I have written this book. My task is to explain why terms such as "missing link" encapsulate more than a century of error in thinking about evolution, particularly of human beings. They reinforce a monstrous view of evolution whose function is to cement our own self-regard as the imagined pinnacle of creation, the acme, alpha, and omega of evolution.[2]

Evolution is a word we use to describe changes in organisms due to the interaction of hereditary variation, superabundance, environmental change, and time. Evolution has neither memory nor foresight. It has no scheme, design, or plan. Now, it might be the case that trends, such as one leading remorselessly and finally to the human state, are apparent in evolution, but these are, by necessity, seen after the fact, and are not built into the process beforehand. The patterns we see in life are the results of evolution, and are contingent. In and of itself, evolution carries no implication of progression or improvement. Absolutely none. Zip. Nada.

The term "missing link," however, speaks to an idea in which evolving organisms are following predestined tracks, like trains chugging along a route in an entirely predictable way. It implies that we can discern the pattern of evolution as something entirely in tune with our expectations, such that a newly found fossil fills a gap that we knew was there from the outset. Quite apart from the impossibility of knowing whether any particular fossil we might find is our ancestor or anyone else's, this is a model of evolution that is at once entirely erroneous, and also rather sad.

In my time as fossil-watcher at *Nature*, the most interest has been sparked by fossils that challenge our expectations, rather than those that confirm them: jolting us out of well-worn mind-sets and forcing us to look at the world in an entirely new way. Fossils such as *Sinosauropteryx*, the first of many dinosaur species announced that had feathers, or feather-like integumentary structures, prompt us to reassess the evolution of birds and flight; fossils of the aquatic, fish-like amphibian *Acanthostega* and the amphibian-like fish *Tiktaalik* compel us to reassess our ideas about how fish evolved legs and left the water; fossils such as *Homo floresiensis*, with its mute assertion of the unexpected yet likely richness of human diversity in the recent past, show us that there is more than one way to be counted as human.

Whatever its position in evolution, *Darwinius* was a living organism worthy of study and respect as a creature in its own right; it did not exist by virtue of being a staging post in the predictable evolution of anything else. For that reason, hailing something—*anything*—as a missing link only cheapens that which we wish to exalt.

In this book I shall show you how and why the view of evolution presented in the popular media is wrong and why we cannot use it to bolster our own position in creation. I shall also show you how to challenge what one reader of a draft of this book has called "human exceptionalism"—the tendency to see human beings as exceptional by virtue of various attributes such as language, technology, or consciousness. There is nothing special about being human, any more than there is anything special about being a guinea pig or a geranium. This insight should allow you see the world afresh, and marvel at each and every creature as it is, for its innate wonder and uniqueness, not as a way station toward some nebulous, imagined transcendence.

The very beginnings of this book lie with two friends, fellow authors and mentors. The first was paleontologist Chris McGowan. I bumped into Chris in the lobby of the Congress Hotel in Chicago in 1996, where we were both attending the North American Paleontological Convention. I was looking for an agent at the time, and, as I admired Chris's books, asked him for a recommendation. He kindly gave me the details of his agent, Jill Grinberg, who represents me to this day.

"You should write a book about human evolution," Chris suggested, helpfully.

"But the World and his Dog have written books about human evolution," I complained—not without justification, as I'd just reviewed a whole stack for the *London Review of Books*.[3] "What the world does *not* need," I went on, "is yet another book on human evolution."

"Ah," Chris responded, "but this would be *your* book on human evolution." This, Chris, is that book. *My* book on human evolution.

The second was the late John Maddox, editor of *Nature*, who in 1987 hired me to work on that august, historic magazine as a junior reporter. I was an unlikely candidate, with virtually no writing experience, and I hadn't yet completed my PhD. But Maddox inexplicably took a shine to me, took me under his wing, and taught me all I know—I owe him an immense debt.

In 1998 Maddox published a book called *What Remains to Be Discovered*.[4] As anyone who read the book and knew its author would instantly realize, the book was characteristic—astonishingly erudite (Maddox really did know *everything*), arch, iconoclastic, exasperating, contrary, and prescient. What set this book apart from the mass (and still does) was Maddox's conviction—distilled from a lifetime in and around science—that the most interesting things about science are not what we know, but what we *don't* know. To go further, it is a fascination with the unknown that motivates scientists. Part of the reason is an intuitive understanding that the more we find out, the more our ignorance grows.

It is a wonder, therefore, that some people—including educators, journalists, and scientists—do not seem to *get* this. To them, science is all about Facts—like the educational program of Mr. Gradgrind in the opening scene of Charles Dickens's *Hard Times*. Facts equate to Truth, and science, they appear to think, is a zero-sum game, all about increasing the quantity of truth and diminishing the net volume of ignorance. In reality, science is about neither Facts nor Truth, but the quantification of doubt. In the small corner of reality that is available to us, scientists set limits on ignorance—but can never banish it entirely. And, to repeat, the more we discover, the more extensive we find the ocean of ignorance. The well-worn response to any new finding—that it "raises more questions than it answers"—is a cliché for good reason. When I go to talk to scientists about the inner workings of *Nature*, I announce—with pride—that everything *Nature* publishes is "wrong." This shouldn't really be as shocking as it is. After all, any answer in science isn't the Last Word, and indeed can never be so. All scientific discoveries are provisional, set to be overturned by results gained from more data, better instrumentation, and new ideas.

The book proposal I initially sent to Jill was called *Dinosaurs Don't Climb Trees*, which morphed into *Thirty Ghosts* and eventually *In Search of Deep Time*, in which I aired the idea that knowledge is not a simple matter of accumulating Facts, but circumscribing more nebulous realms of doubt. I didn't put it in so many words, though—what came out was an exegesis on cladistics, a method of reconstructing evolutionary history that circumvents the assumptions paleontologists and others make about the completeness of the fossil record such that we can reliably read it as a story, in any scientific sense.

That creationists quoted from this book with gusto was perhaps to be expected (I address this issue later on in this book), but what I hadn't expected were howls of indignation from some paleontologists accustomed to using the records of various organisms as vehicles to infer past history.[5]

As Sam Goldwyn once memorably observed, we've passed a lot of water since then. I followed *Deep Time* with a variety of books, from serious pop-science (*Jacob's Ladder*) to fannish criticism (*The Science of Middle-earth*) to a coffee-table book (*A Field Guide to Dinosaurs*, illustrated by the incomparable Luis V. Rey) and even fiction (*By the Sea* and *The Sigil*)—and yet, despite their variety, all seem to draw from the same inspirational spring. That is, that science begins and ends with an appreciation of the unknown, of the vastness of our ignorance, and that this demands not arrogance but humility before the evidence. This is where, I think, the brave souls attempting to stem the creeping tide of willful (often religiously motivated) ignorance have failed. Rather than trumpeting loudly the virtues of Science, Truth, and—yes—"Facts" over Pseudoscience and Superstition, they should admit the obvious.

That is, science is not about Truth, but Doubt; not Knowledge, but Ignorance; not Certainty, but Uncertainty. Never in the field of human inquiry have so many known so much about so little. Only creationists, who are vouchsafed the answers at the back of the book (or, in this context, at the front of The Book), can afford the swaddling comfort and deceptive luxury of Truth, of Knowledge, of Facts that can be Known—because they "know" the answers already, having accepted them without question from a higher authority, as a child from a parent.

Scientists, even those who don't know their scripture,[6] who have grown up, so that they feel capable of looking for their own answers rather than having them handed down to them from above, should be able to convey the wonder—the awe, terror, and insignificance—engendered by confrontation with the unknown. That, really, is what all my books have been about, and this one—I hope—represents a distillation of my entire worldview.

Once upon a time we thought the earth was the center of the universe, but were shocked to find that this was not the case. We thought that Man was the pinnacle of Creation, but despite Darwin, many still cling to this view—for which there is neither any excuse nor justification.

1: *An Unexpected Party*

Many years ago I was a paleontologist. I studied fossil bones. Each bone is mute testimony to the existence of a life, in the past: of an animal the likes of which might have vanished from the earth. I gave up being a full-time bone-botherer when I found myself on the staff of *Nature*, the leading international journal of science.

I was a junior news reporter on a three-month contract. My first assignment, at 9:30 a.m. on Monday, 11 December 1987, was to write a brief piece on new radiological protection guidelines, of which I knew nothing whatsoever. By noon, however, I was to deliver a well-turned, terse, and, most importantly, authoritative story that could stand the scrutiny of *Nature*'s discerning readers.[1]

It wasn't long before I accreted the job of writing *Nature*'s weekly press release—a document that goes out to journalists around the world, keen to learn the latest stories from the frontiers of science. Given that, like me, many journalists would be unlikely to understand all the technical details in each paper, my task was to write a document that would summarize the essence of each in language that would be generally comprehensible. It was an enjoyable and mind-stretching task. On any given day I might be writing about anything in science, from high-energy physics to the molecular biology of HIV-1.

I also got some practice at writing catchy headlines.

My favorite press-release headline concerned a story about mice apt to lose their balance and fall over.[2] The researchers found a genetic mutation responsible for this defect. The research was important because it allowed an insight into a distressing hereditary disease called Usher's syndrome, which is responsible for most cases of deaf-blindness in humans, and which can also include loss of balance. To paraphrase what the humorist Tom Lehrer noted about himself, my muse is sometimes unconstrained by such considerations as taste: so my headline was (hey, you're way ahead of me here)

THE FALL OF THE MOUSE OF USHER

A perk of being the press-release writer was to sit on the weekly meeting of editors trying to decide what would be on *Nature*'s cover two weeks hence. It was here that I first began to appreciate that editors at *Nature* are among the first to hear about new insights into the unknown. In 1994, two marine biologists sent us an amazing photo captured by the Alvin submersible at a depth of more than 2,500 meters. The picture was dramatic, contrasty, and gothic. Picked out in harsh spotlights, exposé-style, it showed two octopi, each of a different species unknown to science, but both male, and having sex.[3] A colleague suggested that this would make an arresting cover picture—another, however, demurred, on the grounds that it was "disgusting." At this point I spoke up—I can still hear myself saying the words—"we can always put black rectangles over their eyes." My mind raced ahead, composing an arresting press-release entry that would be headed with the line

BESTIAL SODOMY IN THE ABYSS

In this case, taste intervened and I used something less lurid. The picture didn't make the cover, either.

I mention all this to excuse some of what follows—if I am critical of journalists and news editors, my criticism comes from experience. I know what it is like to work on a story to a tight deadline, and from a position of relative ignorance. I can also appreciate that the term "missing link," which seems to encapsulate so much in so little space, exerts an almost irresistible allure, even though it represents a completely misleading view of what evolution is, how it works, and the place that human beings occupy in nature.

In the course of time, I migrated from the news department to the "back half," the team of editors who have the immense privilege of selecting which research papers from the stream of submissions will be published in the journal. One of the pleasures of the job is receiving the first news of important, potentially world-changing discoveries.

An account of perhaps the single most remarkable discovery I've seen in my career as an editor was submitted to *Nature* on 3 March 2004. The discovery was of something quite unexpected, opening up unsuspected vistas on things we didn't know we didn't know, and challenging conventional assumptions about the inevitable ascent of humankind to a preordained state as the apotheosis and zenith of all creation. After several revisions, and much discussion among my colleagues and the panel of scientists we'd assembled to advise us on the report of the discovery, the news was published in *Nature* on 28 October 2004.[4]

This communiqué from beyond the realms of the known came from an international team of archaeologists working in a cave called Liang Bua, on the remote island of Flores, in Indonesia. If you want to find Flores on a map, look up the island of Java, and work your way eastward, past Bali and Lombok, and there it is. Flores is part of a long chain of islands that ends up at the island of Timor, well on the way to Australia, New Guinea, and the Pacific Ocean.

One of the more intriguing questions in archaeology is when Australia was first settled by modern humans, the ancestors of today's aboriginal peoples. There is much debate about this issue. Clearly, one way of illuminating the problem is to search for early modern humans living in what is now Indonesia, which can be thought of as a series of stepping-stones between mainland Asia and Australia.[5] That's where Flores comes in. Archaeologists are interested in the caves of Flores and other islands such as Timor because of their potential to yield remains of *Homo sapiens*, modern people caught in the act of heading toward that distant island continent later associated with cold lager, "Waltzing Matilda," and *The Adventures of Priscilla, Queen of the Desert*. This is what drew an international team of archaeologists to Flores, and in particular to Liang Bua, known as an archaeological site for decades.

Flores, though, is an island of mysteries—for it has been inhabited for at least a million years,[6] and not by *Homo sapiens*. Stone tools have been discovered in several places on the island, and their makers are usually thought to have been *Homo erectus*, an earlier hominin,[7] whose remains are well known from Java, China, and other parts of the world. The bones of these early inhabitants of Flores have not been found, their presence betrayed only by the distinctive stone tools they left behind.

But whoever these early inhabitants were, their very presence is a problem. In the depths of the ice ages, when much of the earth's water was locked up in ice caps and glaciers, the sea receded so far that many of the islands of Indonesia were connected by land bridges—they could be colonized by anything able to walk there. Not so Flores: this remained separate, cut off from mainland Asia by a deep channel. *Homo erectus*—if that's who it was—must have made the crossing from the nearest island by boat or raft, or, like other animals, washed up there by accident. Once they made landfall on Flores, there they stayed—cut off from the rest of the world for a very long time.

Isolation on islands does strange things to castaways, making them look very different from their cousins on the mainland. So it was with Flores, home to a species of elephant shrunken to the size of a pony, rats grown to the size of terriers, and gigantic monitor lizards that made modern Komodo dragons look kittenish by comparison.[8]

Such peculiar faunas are typical of islands cut off from the mainland where, for reasons still unclear, small animals evolve to become larger, and large animals evolve to become smaller. Miniature elephants, in particular, were rather common in the ice ages. Practically every isolated island had its own species.[9] The one on Malta lived eye-to-eye with a gigantic species of swan called *Cygnus falconeri*, with a wingspan of around three meters.[10] Micromammoths evolved on Wrangel Island in the Russian Arctic, where they outlived their larger mainland cousins by thousands of years.[11]

The fate of island faunas was an important consideration for Charles Darwin, who marveled at the creatures of the Galápagos Islands in the Pacific Ocean, when HMS *Beagle* visited in 1835. Darwin noted that each island had its own species of giant tortoise, as well as its own finches—different from one another yet plainly similar to finches from the mainland of South America. Had some stray finches, once marooned on the Galápagos, evolved in their own way?

The scene is set, then, for Flores, where, at Liang Bua, archaeologists surrounded by the bizarre sought for something so seemingly prosaic as signs of modern humans.

What they found instead was a skeleton, not of a modern human or anything like one, but a hominin shrunken to no more than a meter in height, with a tiny skull that would have contained a brain no larger than that of a chimpanzee.

In some ways the skull looked disarmingly humanlike. It was round and smooth, just like a human skull, and with no sign of an apelike snout. In other ways it was a throwback. The jaw had no chin—the presence of a chin is a hallmark of modern humans, *Homo sapiens*. The arms, legs, and feet of the creature were most odd, looking less like those of modern humans than those of "Lucy" (*Australopithecus afarensis*), a hominin that lived in Africa more than 3 million years ago. The big surprise, though, was its geological age. Despite its very ancient-looking appearance, the skeleton was dated to around 18,000 years ago. In terms of human evolution, this is an eyeblink, hardly rating as the day before yesterday. By that time, fully modern humans, having evolved in Africa almost 200,000 years ago, had spread throughout much of the Old World. They had long been resident in Indonesia, and indeed, Australia.

So what was this peculiar imp of a creature doing on Flores, seemingly so out of tune with its times?

Despite the tiny brain, the creature seemed to have made tools. Pinning tools on a toolmaker is very hard (we weren't there to see them do it), but these tools looked very like those known to have been made on Flores hundreds of thousands of years earlier, presumably by *Homo erectus*. The only difference was that they were smaller, as if fitted to tiny hands. Had the archaeologists discovered a hitherto unknown species of hominin, dwarfed by long isolation alongside the miniature elephants?

Further work at Liang Bua showed that the first skull and skeleton were no flukes. The skeleton was soon joined by a collection of more fragmentary remains, though no more skulls.[12] All the remains could be attributed to the same species of tiny hominin, and showed its presence at Liang Bua, off and on, from as long ago as 95,000 years ago (well before *Homo sapiens* arrived in the area, as far as we know) to as recently as 12,000 years ago.

After that—catastrophe. A layer of ash found in the upper sediments at Liang Bua indicate that many of the inhabitants of Flores were destroyed in a volcanic eruption around 12,000 years ago. The calamity swept away the fairy-tale fauna of giant lizards, tiny elephants, and tiny people (though the giant rats are still there, to this day). More recent sediments, laid down after the eruption, betray the presence of modern humans, their tools, and their domestic animals.

The account that reached my desk at *Nature* made it plain that the discoverers were as honestly puzzled by their discovery as anyone else would have been, in this coal-face confrontation with the absolutely unknown and unexpected—a new species of hominin that lived until almost historical times, but with a weird, antique anatomy and a very, very small brain indeed.

To emphasize the strangeness of the creature, the discoverers gave it a scientific name that was noncommittal, yet set it apart from anything discovered hitherto. They called it *Sundanthropus florianus*—the Man from Flores, in the Sunda Islands. However, the panel of experts I called on to comment on the draft paper, and to make suggestions for its improvement, pointed out how relatively modern the skull looked—how much it looked like our own genus, *Homo*. One commentator also noted that "florianus" didn't actually mean "from Flores" so much as "flowery anus." Clearly, some revision was required.

When the revised paper was published in October, the creature had become *Homo floresiensis*—Flores Man. The skeleton with its skull was catalogued as LB-1, but the media were quick to catch on to a suggestion of one of the discoverers that it should be known as the "Hobbit," after the diminutive hole-dwellers of J. R. R. Tolkien's fiction—though we in the *Nature* office sometimes referred to her as "Flo" (the skeleton having been described as that of a female).

The paper—and the several commentaries that appeared in its wake—saw the Hobbit as a member of a race of humanlike creatures that had evolved in isolation, on Flores itself or nearby, perhaps descendants of the full-sized toolmakers known to have been on Flores for as long as a million years. If isolation on islands could do strange things to creatures as varied as birds and elephants, lizards and tortoises, there seemed no reason in principle why hominins should be exempt. The Hobbit could easily be seen as a relative of *Homo erectus*, known from remains on mainland Asia to be almost as tall as a modern human—but dwarfed as a result of isolation, alongside the elephants whose island it shared.

And then the fun started.

Hardly had the ink dried on the first account of the Hobbit when the backlash began.[13] Critics were exercised by two particular aspects of the discovery.

First, that such an archaic-looking creature had existed so recently, in a region already long inhabited by modern humans.

Second, that a creature with such an incredibly tiny brain could have made tools. The brain was so tiny, even in proportion to the tiny body, that the Hobbit must—the critics reasoned—have been suffering from a physical or genetic abnormality.

Although criticism of the find came in various shades, critics were united, more or less, in proposing an alternative scenario for the existence of the Hobbit. Rather than it being a distinct species, a relic of an older world preserved out of time, it was a form of modern human suffering from microcephaly, a congenital disorder that produces midgets with abnormally small heads.[14]

The first objection can be seen as a symptom of human exceptionalism, the erroneous yet deeply ingrained tendency that I seek to explode in this book. That is, the tendency to see ourselves as the inevitable culmination of a progressive trend of advancement in evolution. The discovery of such a primitive-looking creature living on the same planet at the same time as *Homo sapiens* challenges that view. It is a perhaps unfortunate fact that the only hominin that still exists on Earth is our own. This fact rather reinforces the idea that various species of hominin—the "missing links"—each more humanlike than the one before, succeeded one another with the planned inevitability of runners in a relay race, and that it is not somehow possible for several species of hominin to coexist on the same planet.

It was not always so. As recently as 50,000 years ago, there were at least four different kinds of hominin on Earth—*Homo sapiens* in Africa, Neanderthals (*Homo neanderthalensis*) in Europe and western Asia, and *Homo erectus* in southeastern Asia, to which must now be added the obscure "Denisovans" from eastern Asia.[15] The addition of a fifth—*Homo floresiensis*—would, in such circumstances, hardly be a surprise: neither should it be a surprise were yet more distinct forms of human to be discovered. Indeed, the only period in which only one species of hominin walks the earth is right now. Modern times are the exception, not the norm.

That different hominins might live together in the same region should, likewise, not be a surprise. It is known that various kinds of early *Homo* coexisted with australopiths in east Africa between 2 and 3 million years ago, and that humans and Neanderthals coexisted in Europe for at least 10,000 years (between around 41,000 and 27,000 years ago). The survival of Neanderthal genes in the modern human population[16] shows that the two species occasionally interbred. There can, therefore, be no objection to *Homo floresiensis* as a distinct species, simply on the basis that modern humans were around at the same time; nor on the basis that *Homo floresiensis* looks too primitive to have survived until modern times. As anachronisms go (what people like to call "living fossils"), the Hobbit is hardly a world-beater. Go tell it to the tuatara of New Zealand, the last relic of a lineage of reptiles distinct from a time before dinosaurs evolved, and hardly changed in its external appearance for 250 million years.[17]

The second objection—that the very small brain of *Homo floresiensis* must have been pathological, a symptom of microcephaly—is likewise flawed, but much more interesting.

Microcephalics have heads that are disproportionately small, even for very small people, such as dwarfs or pygmies. It is important to realize that microcephaly has a number of distinct causes. Microcephaly is not one single disorder. Microcephalics suffer from a variety of other disorders as well as malformations of the skull, face, and limbs, the particular suite of complaints dependent on the variety of microcephaly at issue. Some degree of mental retardation is, perhaps not surprisingly, a feature common to microcephalics in general.

And so it was that the Hobbit was compared with various kinds of microcephalics. However, although the brain of the Hobbit is undoubtedly very small, and the skull and skeleton of LB-1 strange in many ways, its strangeness could not be mapped easily onto any variety of microcephaly recorded for modern humans. That does not mean that the microcephaly idea is ruled out. It could be that LB-1 is the only known exemplar of a hitherto unknown variety of microcephaly. After all, microcephaly of any kind is rather rare, so much so that scientists seeking to compare the Hobbit with microcephalics had to dig deep into the world's medical museums and medical literature even to find the very few specimens of microcephalics available for examination. It is possible that LB-1 suffered from a variety of microcephaly as yet unmapped.

Perhaps the most interesting suggestion of this sort—that LB-1 was a pathological specimen of modern human—was that it was not a microcephalic, but a cretin.[18] Cretinism is not a genetic or inherited disorder, but the result of a chronic deficiency of iodine in the diet. Iodine is a vital component of a hormone, thyroxine, which the body needs for proper growth. Without thyroxine, growth is retarded, and the result is short people, with small heads and various degrees of mental impairment. Iodine is found in seafood, so cretinism is not common close to the sea. It is (or was), however, more common in isolated, inland communities. Liang Bua is in the Floresian hinterland, relatively far from the sea. It is conceivable that LB-1 could have belonged to a tribe of highlanders more prone to cretinism than fisherfolk living on the coast.

But the more that *Homo floresiensis* was studied, especially once the peculiar proportions of its arms and feet became better known, the less well it fit into any known variety of pathology found in modern humans.[19]

The scenarios in which *Homo floresiensis* was not a real species but a pathological version of a modern human were varied, but had one aspect in common: they failed to emphasize (or even mention) that LB-1 wasn't an isolated case that could be singled out as pathological. Remains of the same kind of creature had been recovered from strata at Liang Bua representing an enormous span of time, back to a time before modern humans were known to have existed in the region. This fact alone should have been enough to question any idea that *Homo floresiensis* was a pathological offshoot of modern humans.

The fundamental problem with the microcephaly idea lies less with the idea of microcephaly, or pathology, than that its proponents subscribe to an untenable view of human evolution—one that can only admit to a single pathway of evolution in which human beings stand at the head of a single line of ancestors, each one progressively improved compared with the one before. In that worldview, Flo can *only* be a human being—in which case one then has to explain how she came to look so odd. Proponents of this view tend to be both passionate and argumentative, and become more so as evidence mounts to discredit it. This suggests that the argument is less about one curious fossil than an attempt to shore up a view of the world that is fundamentally mistaken.

The same problem besets the assertion that as a consequence of its small brain, *Homo floresiensis* would not have been able to make tools. It is now known that a wide variety of animals can make tools, many of a sophistication to rival anything made by early hominins. Some of these creatures have very small brains indeed—brain size per se need have little or no connection with technical ability. The idea that brain size matters comes from the view that human evolution is progressive, linear, and inevitably improving.

The problem remained, however, that irrespective of its origins,

Homo floresiensis really did have a disproportionately tiny head. Scaling a modern human down to Hobbit size would have created a creature with a tiny head, but only if it were pathological. The heads of *Homo erectus* were, in contrast, smaller than those of *Homo sapiens*, so perhaps *Homo floresiensis* would be better seen as a dwarfed (but nonpathological) *Homo erectus*. *Homo erectus* was a remarkably variable species with perhaps a tendency to smallness,[20] something that might play in its favor as a possible ancestor of *Homo floresiensis*. Specimens found in the Republic of Georgia dated to around 1.7 million years ago seem to represent a sample of *Homo erectus* of a primitive, early kind.[21] These creatures were small, some comparable in size with *Homo floresiensis*, but their brains were at least twice the size of LB-1. Shrinking *Homo erectus* down to the size of *Homo floresiensis* would still produce a creature with too large a brain. Flo had to have evolved from something smaller still.

Two possible solutions presented themselves. One was a study on island dwarfism in now-extinct hippopotamuses that lived on Madagascar, showing that in some cases, the brains of animals subject to island dwarfism would be reduced more than one would expect, even when one scaled a full-sized animal down to midget size.[22] This makes sense in terms of energetics. A possible cause of island dwarfism is that castaways evolve a smaller size in response to the pressure of reduced resources. The brain is, proverbially, the most expensive organ to run in terms of its mass, and so might be expected to evolve a disproportionately small size. Yet such a reduction has its limits. A brain can't reduce to the extent that function would be impaired. However you look at it, a race of cretins or microcephalics isn't going to survive for very long. But even when the further downsizing of brains of island species was accounted for, the brains of *Homo floresiensis* looked too small, even for *Homo erectus*.

The second solution was that *Homo floresiensis* was a dwarfed version of an even earlier, more primitive hominin than *Homo erectus*, perhaps a creature so primitive that it would not be grouped within the genus *Homo*. This had indeed been an option favored by the original discovery team, but they had to some extent been dissuaded by the panel of experts I'd assembled to assess the original report, who had looked at the skull of LB-1 and said that despite its size it fit better within *Homo* rather than outside it.

After all, what choice was there? It seemed far simpler to admit a new member to our own select genus, no matter how weird the entrant, than to defy everything we thought we knew we knew about the human story: to open the floodgates of the unknown unknown.

Conventional wisdom suggests that the entire tale of human evolution had taken place exclusively in Africa until around 1.8 million years ago, when *Homo erectus* became the first hominin to leave that continent and colonize much of the rest of the Old World. The fossils from Georgia might have represented this wave of emigration. There is no compelling evidence that earlier hominins, such as *Australopithecus* or the earliest members of our own genus such as *Homo habilis* (somewhat more like *Australopithecus* than *Homo erectus* in many ways) had ever left Africa.[23] *Homo floresiensis* just might be that first piece of evidence.

Perhaps some hominin left Africa long before *Homo erectus* had evolved, migrating across the Old World, evolving into all sorts of diverse and unimagined forms, the only trace of such an adventure being a single, late-surviving relic marooned on remote and distant Flores.

When the researchers unearthed *Homo floresiensis* from its long home, they opened the door to things we not only didn't know, but didn't even suspect, so wedded were we to the canonical out-of-Africa picture: not just to a remarkable, almost unbelievable testament to the power of evolution to shape living matter into unexpected shapes; but to a hitherto unknown and unsuspected chapter in human evolution, a vista far greater and more varied than anyone had dreamed possible.

I have chosen to highlight the case of *Homo floresiensis* as it's the best example I can think of, from my own experience, of a new discovery that challenges our expectations, our restricted notions of evolution based on human exceptionalism, and with it an idea of progressive improvement.

The tale of the Hobbit is the book in microcosm. It shows that new discoveries often challenge deeply held notions of how we think evolution really ought to have happened, such that we humans are the culmination of some cosmic striving for order and perfection. It also shows us that stories we tell based on fossils are often easily bruised by the sheer scale of our ignorance. Fossilization is rare—so rare that there could well have been an entire episode of human evolution, a pre-*Homo* exodus from Africa, that has left no trace in the geological record other than the Hobbit.

If there is one lesson that science holds for us, it is this—that our special estate, based either on a progressive scheme of evolution leading to its inevitable human culmination, or on a narrative reading of prehistory, is never justified. It was Charles Darwin himself who put it best. Right at the end of *The Origin of Species*, he presented the idea of the "tangled bank," his vision of evolution in action: "It is interesting to contemplate a tangled bank, clothed with many plants of many kinds, with birds singing on the bushes, with various insects flitting about, and with worms crawling through the damp earth, and to reflect that these elaborately constructed forms, so different from each other, and dependent upon each other in so complex a manner, have all been produced by laws acting around us." From this, it's evident that Darwin saw evolution not as progressive or improving, but as an activity that happens in the continuous present, as creatures interact with one another, moment by moment. From this it is clear that evolution has no plan. It has neither memory nor foresight. No vestige of cosmic strivings from some remote beginning; no prospect of revelatory culmination in some transcendent end.

Rather than being at the pinnacle of creation, human beings are just one species on the tangled bank of Darwin's imagination. Human beings are special in many ways—of course we are—but so is each and every other species, from the insects flitting above the bank to the birds perching on the branches to the worms struggling through the damp earth beneath.

The idea of progress is, however, deeply pervasive. Our culture is drenched in it: our politics, our economics, and our science, including (and perhaps especially) evolution. It is always assumed that things advance unerringly upward as if motivated by some inherent force. Progress is unstoppable. What's more, we are told that we need it, that we are reliant on it, and that the stagnation or reversal of progress is a Bad Thing.

Progress is destiny. The only way is up.

To be sure, if you look back from the viewpoint of the present day to any period in the past, progress seems natural and inevitable. But such a perspective is limited, because it denies that any other course might have been possible, and edits out any promising side branches that went nowhere.

An important concept to take away from this discussion is that of *loss*. Stories of progress are written by history's victors—or at least its survivors. Such tales tend to talk of increasing complexity and sophistication, and ignore the perhaps different perspectives of creatures that have become extinct.

The concept of loss is vital to a proper understanding of evolution. This is especially so for human evolution, a subject that is often deficient in perspective: understandably so, because we, the storytellers, are only human. The history of life told by other organisms might have different priorities. Giraffe scientists would no doubt write of evolutionary progress in terms of lengthening necks, rather than larger brains or toolmaking skill. So much for human superiority. If that's not ignominy enough, bacterial scientists would no doubt ignore humans completely except as convenient habitats, the passive scenery against which the bacterial drama is cast. Now, ask yourself—which of these stories is any more valid than any other, at least as a narrative?

The late Stephen Jay Gould punctured the idea of inevitable progression in his book *Wonderful Life*, by introducing the concept of "contingency." That is, creatures need to be more than fitted to their lives and lifestyles by evolution: they also need a generous dollop of luck. Once luck has been stirred in, the whole idea of progress driven by some innate striving, or superiority, or destiny, becomes nonsense.

Gould started *Wonderful Life* by showing how our idea of human evolution as a matter of inevitable progress is so deeply ingrained in our culture that admen use it as a way to sell products. Admen use the metaphor of human evolution so frequently that it's become a cliché. You'll no doubt have seen a progression of apelike beings, walking from left to right, each one following the next, each more upright and humanlike than the last. Figure 1 is my own modest contribution to the canon.

Figure 1

Admen complete this familiar parade with the latest computer or washing machine. The subtext is that the consumer product we're being urged to buy is the result of successive improvements in a kind of mechanical evolution, each better than the one before. Some commercials even exploit popular notions of evolution explicitly. The TV commercial that presents evolution as a device to produce a creature sufficiently evolved to appreciate Guinness beer was especially memorable.

My favorite variation on this theme concerns a car. "It's Evolved" purrs the voice-over.[24]

The idea of human evolution as a tale of inevitable progress is, however, a travesty, and has nothing much to do with Darwin. The bastardized view of evolution that's become so much a part of the general consciousness—so much so that it's so much low-hanging fruit for admen—owes much to Ernst Haeckel, Darwin's number one fan in Germany.

Haeckel took Darwinian natural selection and bolted on to it older ideas of progress popular among nineteenth-century German thinkers. Take a series of forms, each one more advanced than the last—according to whatever criterion you desire, be it larger brains, longer necks, more prominent plumage, whatever—and simply draw arrows between them, representing some innate striving toward cosmic perfection.

And that's evolution—or, at least, evolution as most people think of it—a kind of cartoon, infused more by our prejudices, desires, and innate self-regard than any actual evidence. Figure 2 is my version—my parade of likely characters, linked by arrows, pointing in the direction of progress.

Figure 2

The arrows represent natural selection, or evolution, as essentially and inherently an agency of inevitable progression, with—perhaps—the aim of producing, in its final form, the perfection that is Man (with a capital *M*).

Yes, to be sure, I'm having a lot of fun at everyone's expense, but how can this cartoon be in any way wrong as a picture of human ancestry, at least in a general sense?

No sensible, informed person would doubt that we all had ancestors, and the further back you look in time, the more apelike they'd have looked, right?

This cartoon picture of human evolution doesn't really represent our actual ancestors, but metaphors, right?

Well, yes, up to a point. That we all had ancestors is true—emphatically so. That our ancestors would have been more apelike the further back in time we look is also highly likely, given that our closest living relative in the animal kingdom is the chimpanzee, and it remains reasonable to suggest that the chimpanzee has evolved less far from our common ancestral state than we have.[25]

But if the figures in this parade are metaphors, what do these metaphors represent? The figures, surely, represent idealized evolutionary stages, between ancient ape and modern human, rather than specific individuals or even species. That's fair enough.

But my argument is less with the figures themselves than the arrows between them, arrows that seem to represent inevitability and progress, of evolution leading, inexorably, through one lineage and one lineage alone, to its culmination, the latest model human (or washing machine or car), more refined, more sophisticated, and crucially, more perfect than the one before.

Another problem pointed out by Darwin in the *Origin* was what he called the "imperfection" of the fossil record. The record of life preserved as fossils is immediate evidence for evolution having happened. It is, however, rarely good enough for us to be able to trace the evolution of one particular species from another with any confidence. It is important to remember that fossils, on their own, are remnants of creatures toiling on some tangled bank of the past. They do not, of themselves, represent coherent statements about evolutionary history—still less, evolutionary progress. If a fossil is a statement, it is not a sentence, such as

Here lieth ye ancestor of all humans,

because fossils are not buried with their pedigrees, nor prognostications on the future of their progeny, if any. No, fossils are not statements. Nor are they phrases, or words, but exclamations, from which we, the finders, are invited to make what we can.

Piecing together the tale of evolution from fragmentary fossils is a hard business. Because fossils are so rare, and because an unknowably large proportion of the history of any lineage will have been erased, what fossil hunters can never do with confidence is look at a fossil and assert that it is the actual ancestor of any creature now living (or of any other fossil). To be sure, the fossil might be such an ancestor—because, after all, we do have ancestors—but the chances of this in any particular case are unknowable, and in any case vanishingly small. There is, happily, a way out of this apparently blind alley, and that is the fact of evolution.

Evolution is demonstrated by the existence of fossils and the community of all life. By this, I mean that the chemistry that animates you is virtually identical to that which animates every other living creature. Because of this, there is very good reason to suspect that all life shares a single common ancestor.[26] This is more than a supposition—the notion of a single common ancestry has been tested, formally and rigorously, and has been found to support the pattern of extant life far better than any model positing independent origins.[27] It follows, therefore, that any fossil we find will be a cousin, in some degree, of any other creature, living or extinct, discovered or undiscovered—even if we can never show that anyone was anyone else's ancestor.

In short, this approach to reconstructing the story of evolution as a matter of degrees of relationship, which can be inferred and tested, is far superior as a scientific approach to evolution than suppositions about of ancestors, descendants, and "missing links"—which can be inferred but never falsified. My task here, though, is to show how the sparseness of the fossil record is sufficient to mislead us, were we bent on thinking of evolution as an onward march of progress and improvement.

Let's try this thought experiment. Imagine that by some divine grace (because you'd get it no other way) you were granted knowledge of the complete history of every individual creature that ever lived—its offspring, its ancestors and descendants—and could draw the true tree of life, that is, what actually happened. Just to make things simpler (and more relevant to our current concerns), you restrict yourself to drawing the true tree of hominin evolution, back to our common ancestor with chimpanzees. I've drawn what this might look like in figure 3.

Time moves from left to right—the left is long ago, the right is more recent. This, the true tree, is very bushy, as you can see, and most of the branches lead nowhere—that is, to extinction. At first glance it's impossible to select any one branch as especially important.

But because of your divinely granted complete knowledge, you'd be able to pick out the line that leads, uniquely, to modern humans. Figure 4 shows the "true tree" with the ancestry of modern humans indicated by a thick line.

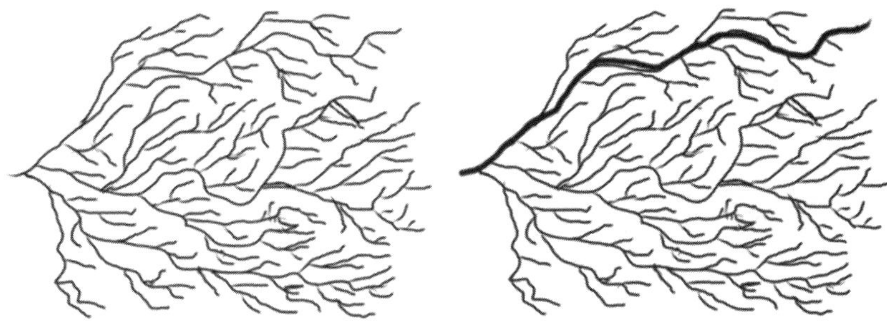

Figure 3 **Figure 4**

It's important to realize that in reality you could have no absolutely certain knowledge of the true line of human evolution—what really happened so surely that you'd *know* you knew it—unless you had a record of every single hominin that ever existed, and full details of their ancestry back to our last common ancestor with chimpanzees.

Back in the real world, we are left with what few scraps time and chance have left us, and that's very few indeed. Primatologist Robert D. Martin estimated that we knew perhaps as much as seven out of every hundred primate species that have ever existed, given a few assumptions about the known diversity of fossil primates, and the number of primate species currently living. Martin made his estimate twenty years ago.[28] Given that the amount of ignorance expands with the gain of knowledge (that the more scientists discover, the more they know that they don't know),[29] that proportion might well have decreased, even though paleontologists have discovered quite a few extinct primate species since then.

In figure 5 I show what remains of the "true tree" once the majority of its branches and twigs have been pruned, leaving only a few fossils. The deletions are, I confess, not totally random. I have been particularly careful to erase any branching points, as fossils will never come complete with that information.

From this scatter, one could arrange many different sequences of fossils from older to younger, and suggest that this sequence might represent a probable evolutionary sequence between apelike ancestor and modern humanity. However, given the evidence at hand, you could link up more or less any sequence of fossils from this scatter and make other assertions of equal validity—without divine grace, who would know which was more likely to be correct?

Figure 5

In figure 6 I show three of the very large number of possible trajectories between ancestor (left) and human (right). All use the same scatter of fossils, but the trajectories are different from one another.

The first is closest to the "truth," but we could never "know" this. The three plots would, I think, have something in common and that's this: the fossils used as links in the chain would, when arranged in the order the arranger assumes to be correct, show a progressive increase in those features assumed (retrospectively) to be characteristic of humans—more erect posture, larger brain, and so on. At no point would there be a reversal, such that a descendant would be more stooped than an ancestor, or have a smaller brain. Not that such things are not possible—the case of the Hobbit shows that they are—but because the assumption of progress is so ingrained that it would not occur to most people that there might be any other course besides onward and upward.

Perhaps the most important thing to take away from this chapter is that new discoveries challenge our idea of progress—that matters are subject to a continual improvement, the refinement of each stage building on that of the one before in seamless progression. What the conceit of progress tends to ignore is the idea of loss—that many experiments in life were made that subsequently went extinct, and so are left out of the canonical tale of improvement.

More than this, the idea of progress tends to be based on criteria that we decide after the fact according to our prejudices, and which need not be the most important or relevant ones. Because we seem to have larger brains and a more erect gait than earlier essays in the human condition, we always assume that the evolution of humanity is a story that must be told in terms of progressive increases in brain size and stature. This reasoning, however, is circular. We have larger brains than our presumed ancestors, so evolution must be couched in terms

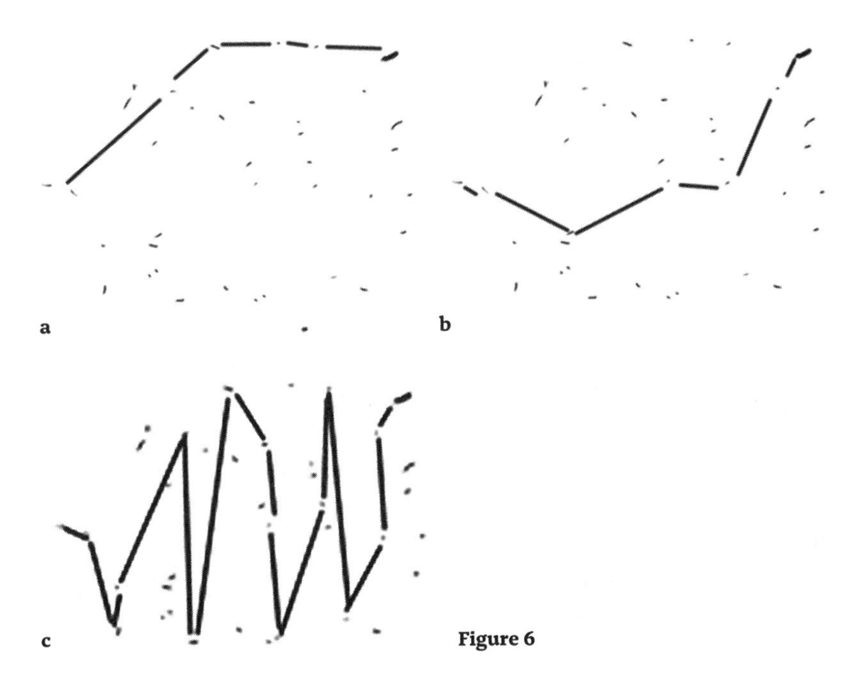

Figure 6

of brain size, so the discovery of creatures living in the past that had smaller brains will naturally confirm our prejudices. For all we know, our picture of human evolution might be better told in terms of, say, changes in the number of kinds of bacteria that live in our small intestines. After all, your body probably contains around ten times as many bacterial cells as human ones.[30]

To really get a grip on why evolutionary arguments about human exceptionalism are wrong, you need to have a good understanding of what evolution is—and what it is *not*.

The next chapters offer a handy cut-out-and-keep guide to evolution by natural selection. You might be surprised to learn that evolution by natural selection is far less—and far more—than you thought it was. After that I'll discuss the concept of loss in more detail, showing how the stories on which we base our fragile suppositions about human exceptionalism are based on very little evidence at all.

2: *All about Evolution*

The word "evolution" is probably one of the most abused words in any argument about science. To some, it is a rallying cry to rationality. To others, it's a term of abuse, the term "evolutionist" hardly less derogatory than "abortionist." There can be few other words that get so much mileage while remaining so poorly understood. "When *I* use a word," said Humpty Dumpty in Lewis Carroll's *Through the Looking-Glass*, "it means just what I choose it to mean—neither more nor less." Matters are made worse by the fact that the meaning of the word has changed over time, and remains ambiguous to this day.

When inventing the wheel, it is best to ensure that it is round before deciding what color to paint it. So, before we can get a handle on the word "evolution" in all its protean and subtle variety, one must first understand how it works, on the most basic nuts-and-bolts level. This is why Darwin started *The Origin of Species* by outlining such a mechanism—and not mentioning the word "evolution" at all. Darwin had very good reasons for not using the word in his masterpiece, as I shall explain a bit further on. Until then one might do a lot worse than follow his example.

Like many people these days, we in the Gee household keep chickens in our backyard. The hens are of several different breeds. We started with bantams, small birds whose function is more ornamental than anything else. They don't lay many eggs, perhaps ninety per bird per year. They are, however, long-lived. At the time of writing, one of our first hens, a Pekin bantam, is four years old and still going strong. Our next two hens, Polish bantams, are almost as old, and in rude and squawking health. We also have several standard-sized hens, which lay more and bigger eggs.

But the prizes for productivity go to those in the flock that started their careers in intensive egg-production facilities. A battery hen can lay as many as three hundred eggs per year, but at a cost—the hens don't live long. When a battery hen stops laying regularly, she dies of old age. Battery hens have been bred that way, to invest as much energy as possible into producing eggs, at a cost to their own bodily maintenance. Our first four battery rescues died of old age within two years, and we are now on our second quartet.

All the battery hens have russet feathers and red combs. They look just like the Rhode Island Reds my mother kept when I was a boy. As every backyard farmer knows, Rhode Islands are just about the best hens to keep if you like lots of eggs. These battery birds plainly have Rhode Island in their heritage, but they've been turbocharged to ramp up egg production at the cost of virtually everything else. In other words, they have been selected. If farmers depend for their livelihood on selling as many eggs as possible, they will breed future stock from the most productive egg-layers, and make the rest of the hens into cat food. They'd continually breed from the best layers in each generation, until, many generations down the line, they'd have created a new breed of hen that routinely lays many more eggs than any hen in the original flock.

This idea—the "artificial" selection by stockmen intent on breeding hens that lay more eggs, sheep with fleecier fleece, bulls with beefier beef, and so on—is intuitive, makes sense to anybody—and was how Darwin started the *Origin*.

What Darwin did next was a master stroke. Once he'd established artificial selection as an obvious and unarguable phenomenon, Darwin used it as an analogy for what goes on in the natural world. In nature the role of farmers is played by the environment. Creatures won't be "artificially" selected by farmers for this trait or that, but "naturally" selected by the ever-changing environmental conditions in which they live. If the climate turns cold, those elephants that happen to have more body hair will be more likely to survive than those that are less hirsute—long enough to breed and pass on their hairiness to their offspring, while the baldies devote their energies to keeping warm rather than reproducing. If the climate continues cold, the bald elephants will eventually be replaced by woolly mammoths.

The beautiful thing about natural selection is its simplicity. All it requires to work are four things, three of which are readily apparent with eyes to see. They are heritable variation, the ever-changing environment, superabundance of offspring, and the passage of long periods of time.

Let's look first at heritable variation. This means that any group of creatures will differ in their appearance or constitutions from one another, and that this variation is inherited from their parents. Unless they are identical siblings, the children in a family will inherit different traits from their parents, to different degrees. Some will be taller, some shorter, some darker, some fairer. For example, if you gathered every adult male (or adult female) in your town and measured them, you'd find that they'd vary greatly in height. You'd have to group men and women separately, as height is in part related to gender—on average, the men in any given population are taller than women from the same population. You'd find that most people would be middling in height, somewhere between 1.5 and 1.9 meters tall. People much shorter or taller than this are relatively rare. Any population is varied, but variation tends to cluster around a "mean" or "average" value. Calculating an average value is easy: add all the heights together, and divide what you get by the number of people you've measured.

The more people you measure, the better, because your result will be a better approximation of reality. If you can't measure everyone in your neighborhood, say, you should still try to measure as large a sample as possible. If you can't do that, you should try to ensure that the people you measure are picked at random. For example, if you measured the heights of the first three people you met, and they happened to be a coven of very small witches, or from a team of very tall basketball players, you shouldn't be surprised that your sample is unrepresentative of people in your neighborhood in general.

When you see reports of preference in the press, such as peoples' voting intentions, or whether their cats prefer ex-battery chicken of one brand over another, you should look out for the small print saying that the evidence comes from a poll of, say, 1,000 people chosen at random. It's important to get lots of people, and to pick them by chance. This chance element is vitally important. There's the probably apocryphal story of a market researcher who found that ninety-nine of a hundred people asked ate porridge for breakfast: it turned out that the people asked all came from the McPherson page of the Inverness telephone directory. This, without meaning any offense to residents of the fine city of Inverness who happen to be called McPherson, is probably not a representative sample of people as a whole.

From this it is clear that variation acts at different levels. As people vary in height even in your neighborhood, so do people from different places. Different populations have different average heights. The average American man is 1.76 meters tall, whereas the average American woman is 1.62 meters tall.[1] Dutch men and women tend to be taller, on average—1.87 and 1.69 meters respectively,[2] whereas urban men and women of the east African nation of Malawi tend to be shorter, 1.67 and 1.55 meters.[3] This means that although men tend to be taller than women in general, the average Dutch woman will be taller than the average Malawian man. Because people tend to marry within their locality or ethnic group, the figures for average height differ from place to place.

Although people vary in all sorts of ways, and even though traits might be influenced by other things, such as nutrition and the environment, it's plain that height tends to run in families—that is, variation is inherited. Tall parents tend to have tall children. My own daughters are among the tallest in their year groups—but I am relatively tall for an Englishman (1.83 meters, against the average of 1.75), and my wife is very much taller than the average Englishwoman (1.8 against 1.6 meters).[4] She also comes from a family of tall women, who tended to marry guardsmen—not just tall, but proverbially tall. Hmm. The tallness strong within them it is.

From all this it's clear that people (and other animals) vary, and that this variation can be passed on through the generations. If this weren't true, then farmers wouldn't be able to breed prime egg-laying hens by selecting the best layers in each generation as brood stock. Such variation is entirely obvious to anybody, yet in Darwin's day nobody knew how variation was maintained. In his time it was generally assumed that the traits of parents got merged among the offspring—but if this were the case, all the variation would quickly get mixed together (like mixing paint of lots of different colors to get brown), and everyone would tend to look the same. But this doesn't happen. Offspring are always varied. Even if the human population were well mixed, such that every person on Earth were obliged to choose their partner through a worldwide dating service, and did so for generations, their children would still vary in height, skin tone, eye color, and a host of other traits. The answer came long after Darwin, with the discovery of genetics, in which it is shown that traits are the expressions of atoms of inheritance called genes, which combine and recombine with one another to create variation, but remain individual and distinct. Some traits are influenced by single genes. Others, such as height, are influenced by many thousands.

The second factor that contributes to natural selection is the variability of the environment in which organisms live. I mentioned the case of mammoths above. If the climate turns cold, hairier elephants will have a better chance of surviving to reproductive age than elephants that are less hairy. Because hairiness will be to some extent inherited, the tendency toward hairiness will spread, so that, over time, the population of elephants will become hairier, on average.

You'll of course have appreciated that the environment is very much more complicated than this cartoon explanation implies. The term "environment" means any circumstance, however small, that affects the chances of a creature surviving long enough to pass its traits on to the next generation. The environment doesn't just mean the climate, or even the weather, but also the relationships that a creature has with other creatures, whether of different species or its own. The environment is therefore not one single thing, but uncountably many, each one changing minute by minute. A creature will have to be able to gather enough resources to grow, all the while trying not to be eaten by other creatures. Once mature, a creature will have to find a mate, and produce offspring, whose interests might differ from its own. All such factors constitute the environment.

Not surprisingly, some parts of the environment actually act in opposition to one another. Perhaps the best-known example is the case of sickle-cell anemia. This is an inherited disorder in which a person's red blood cells fold up like squashed footballs and become very stiff. This makes them poor at carrying oxygen round the body. The malformed cells are also prone to clogging up blood vessels, causing all kinds of potentially life-threatening complications, including increased incidence of infection, damage to internal organs, thrombosis, and stroke. Sickle-cell anemia is a very serious disease indeed, and children with the disease stand much less chance of living long enough to reproduce than children without it. As a result, sickle-cell anemia is rare in most populations—people die of it before they can grow up to have children themselves.

The inheritance of sickling is well understood: it results from a defect in a single gene that codes for part of the molecule of hemoglobin, the protein in red blood cells that carries oxygen in the blood. Most genes are carried in two versions or "alleles," one inherited from the father, the other from the mother. A child can carry two normal alleles, one normal allele alongside one sickling allele, or two sickling alleles. Only that child whose unhappy lot it is to carry two sickling alleles will suffer full-blown anemia. People with two normal alleles will, of course, not get the disease. People with one normal and one sickling allele will be normal, because the normal allele will produce more than enough normal hemoglobin to get by, and they are likely to suffer only if they happen to find themselves up a mountain where oxygen is scarce and hemoglobin has to work overtime.

Now, you'd think that because of the sickling allele's effects on the chances of a young person's reaching adulthood, natural selection would have expunged it pretty smartly from the population. But there's a catch. It so happens that people with the sickle-cell trait are more resistant to malaria than those without. Malaria is debilitating enough in adults, but in children it can be lethal. It is caused by a microscopic parasite that hides out in red blood cells for part of its life cycle. Fewer red blood cells mean a less friendly place for malaria. People with sickle-cell anemia will be very ill anyway, but in the lottery of life, serious illness is often preferable to immediate death. People who have one sickling allele and one normal allele will be very much less ill, but much more resistant to malaria than those with normal alleles.

In parts of the world where malaria is endemic, such as sub-Saharan Africa, a child with sickle-cell anemia, or even a "carrier" with one copy of the sickling allele covered by a normal copy, will be better able to resist malaria and survive than a child with two copies of the normal allele, who is more likely to die from malaria than from sickle-cell anemia. This difference is crucial, for it alters the balance of survival in favor of the child who has sickle-cell anemia over the child who has not—and has allowed the otherwise entirely unwelcome sickle-cell trait to persist. In places haunted by the specter of malaria, carrying a gene for a debilitating disease is actually an advantage—it is the lesser of two evils.

Sickle-cell anemia demonstrates that natural selection is not some agent that drives creatures ever closer to the perfection imagined by advertising copywriters. Far from striving for bigger, better, more complex, or more enlightened, it does *precisely* and *only* what it needs to do to get a creature from egg to adulthood—*and no more*. This can mean carrying a trait for a dreadful disease that happens to offer protection from something worse. And because the environment is complicated, subtle, and ever changing, it is always a mistake to reduce natural selection to a simple mechanism that creates trends or tendencies that can be easily identified as such, and whose causes can easily be worked out.

The third factor that contributes to natural selection is superabundance of offspring. This means that creatures tend to produce many more offspring than can possibly survive. And by "many more," I mean *vastly* more. Anyone who thinks evolution is all about elegance and orderly perfection in nature would be shocked by its profligacy and waste.[5] Next to our chicken run is a pond, which I dug specifically to encourage the arrival of frogs, which would feast on garden pests such as slugs. Each spring the pond bubbles with hot frog-on-frog action, after which the water seethes with thousands of tadpoles—only one or two of which will survive long enough to reach sexual maturity. In the fall, our apple tree is groaning under the weight of fruit, but few or none of its seeds will ever germinate. Every woman produces hundreds of eggs throughout her lifetime, but only a few will be fertilized and come to term; every man produces millions of sperm, but relatively few children.

In ages past, people used to have large families, expecting that many (or most) of their offspring would die of something or another before they reached adulthood. Demons hovered around every crib and outside every nursery. I mentioned malaria, but even today millions of people, most of them children, die from dysentery, diarrhea, tuberculosis, cholera, or the effects of malnutrition. Darwin's daughter Annie died from scarlet fever, which is now relatively rare. When I was a child, less than half a century ago, children even in Britain were severely disabled by or even died from diseases such as measles, mumps, rubella, pertussis (whooping cough), diphtheria, and poliomyelitis. Smallpox was a vanishing threat, but had not at that time been entirely eradicated. There is a reason that many of these dread diseases are associated with childhood—people who contract them as children might not survive to adulthood.

Thanks to improvements in public health and, notably, the success of vaccination, most of these diseases now figure only in period dramas, despite the best efforts of a deluded few anti-vaccination campaigners to turn fiction back into documentary. In the developed world nowadays, mortality among children is less likely to result from infectious disease than from accidents or relatively rare birth defects.

Inherited diseases (as opposed to infectious ones) result from the fact that in a process as complicated and delicate as the development of a creature from an egg, mistakes are often made. The process is so complicated that it's a wonder any of us actually gets born, and it could be that genetic variation itself exists as a hedge against error. By this, I meant that a certain amount of sloppiness is tolerated in the system, creating variation, and those variations that cause lethal or severe inherited disease are the price we all pay for being born at all.[6]

In the meantime—and it sounds desperately cruel—natural selection is likely to favor an earlier death (rather than a later one) from a debilitating disease so that harmful traits are less likely to be passed on (unless they provide an advantage, as in the case of sickle-cell anemia) and, more immediately, so that parents can get on with devoting limited resources to producing healthier offspring instead. In a world in which the threat of disease or mishap is always present, superabundance is a way of beating the odds, of maximizing your chances of your progeny surviving long enough to reproduce. The gambler at the roulette table who places all his chips on a single outcome will almost certainly lose. The gambler who puts a chip on every possible outcome is bound to win something. The second gambler will have lost an awful lot of chips but can stay in the game, whereas the first will have lost all of them and has no choice but to leave the casino.

These three things—heritable variation, the changing environment, and superabundance of offspring—are neither particularly special nor inherently mysterious. The fourth factor is time, and that's a little more tricky.

Darwin saw natural selection not as an agency in itself, but the ongoing result of the interaction of several factors. Creatures tend to produce offspring that vary, and this variation is heritable. They also tend to produce more of them than can possibly survive. Nature will select those few offspring that are most suited to living in the prevailing environment, in much the same way that a stockman will select those animals most suited to his ends. Given enough time, the creatures will change, their adaptations tracking changes in the environment.

But how much time is "enough"? Darwin envisaged that change would be slow, perhaps even imperceptible on the scale of human lifetimes, and reasoned that many millions of years would be required for natural selection to transform a blob of primordial protoplasm into the diversity of animals and plants we see all around us. The problem was that, in Darwin's youth, such time didn't exist. So no matter how obvious heritable variation, superabundance, and environmental change are to every child and countryman, without time, natural selection wouldn't be able to do very much.

What do I mean by time "not existing"? I'm being deliberately arch here. Nowadays we are accustomed to thinking of the earth as very old—around 4,500,000,000 years old, in fact—plenty of time for natural selection to have done its work. We are inclined to take such things for granted, so it's very hard for us to put ourselves into the minds of the average Victorian who had no reason to doubt that the earth was any more than the 5,500 years or so required by the Bible. It took quite a long time for even those interested in the subject to realize that the earth is very much older than this, and even then, only when they were confronted by an otherwise insupportable weight of evidence. (The many people who to this day cling to the old biblical timescale have no such excuse.)

And that's it. Take heritable variation, the changeable environment, superabundance, and time. All these things can be seen—or, at least, understood—by anyone.

So much for natural selection. What, then, about evolution? How is one related to the other? The terms are not equivalent, and that's part of the problem. Here I hope to disentangle the word from some of its ancient baggage, look into its history as a word as well as a concept, and show what (I think) Darwin meant it to mean—which is (I think) rather different from what most people think when they use the term. In fact, I'd go as far as saying that it would be hard to find a worse choice of word than "evolution" to describe what Darwin, very sensibly, called "descent with modification." To Darwin, the word "evolution" did not mean what we think it means today.

As you might expect, the word has Latin roots. According to the online *Oxford English Dictionary*, henceforth *OED*,[7] the Roman writer Cicero used *evolutio* to mean the action of unrolling a scroll. Thus was born the concept of evolution as a process of development, elaboration, and, with it, revelation—that is, the deliberate transformation, by the action of unrolling, of a closed scroll to an open one whose information might be read: an orderly dance from simplicity into complexity. Medieval Latin texts use the term to refer to the passage of time during which any metaphorical unrolling might take place.

The first recorded use of the word "evolution" in English was in 1616, in a translation of the *Tactics* by the second-century Greek military historian Aelian (Aelianus Tacticus), where it means, quite specifically, the movement of forces from one position to another:

> The nature of this Euolution is clearely to leaue the File-leaders in front, and Bringers-vp in reare.

This nuanced view of evolution, as a series of maneuvers along a studied course from known beginning to desired conclusion, broadened to describe the occult movements of the wands of wizards, the gyrations of gymnasts, and, eventually, the choreography of dancers. The many examples given by the *OED* have one thing in common—that the term "evolution" in this sense came to encapsulate an exact, directed and predetermined series of events, as predetermined as a choreographed dance routine. More generally, the word "evolution" came to mean the opening out or unfolding of a series of events in an orderly succession, or the action of elaborating a simple idea into something more rounded, very much by analogy with Cicero's unfurling scroll. As an aside, almost, consider this notable example from Erasmus Darwin's *Zoonomia* (1801):

> The world . . . might have been gradually produced from very small beginnings . . . rather than by a sudden evolution of the whole by the Almighty fiat.

Given what we think we know of evolution—as a gradual process—it is startling to come across Charles Darwin's grandfather Erasmus using the term in precisely the opposite sense.

Those admen I lampooned in chapter 1 would find in the *OED* plenty of precedents for their use of the term "evolution" to refer to the refinement of consumer products (the first recorded such usage being in 1882). But in biology, as in life more generally, the term began to be used very much by way of analogy with Cicero's original meaning—the elaboration of something simple into something more complex, such as a plant from a germinating seed, or the development of a butterfly from a caterpillar—like so many scrolls unrolling, each in its own precise, preprogrammed manner. Here is an entry from the earliest days of the *Philosophical Transactions of the Royal Society*, in 1670:

> By the word Change is nothing else to be understood but a gradual and natural Evolution and Growth of the parts.

And once again from Erasmus Darwin:

> The gradual evolution of the young animal or plant from its egg or seed.

As a term, evolution gets around. I haven't mentioned the several different usages of "evolution" in mathematics, astronomy, and chemistry. All of the above, of course, is by way of a curtain-raiser to what the *OED* lists as sense 8 of evolution (out of eleven), namely "the transformation of animals, plants and other living organisms into different forms by the accumulation of changes over successive generations." The first recorded use of "evolution" in this sense is in 1832, in Charles Lyell's *Principles of Geology*, a work with which Charles Darwin was very familiar.

> The testacea of the ocean existed first, until some of them by gradual evolution, were improved into those inhabiting the land.

As I noted, Darwin did not use the word "evolution" in the *Origin* (and continued not to do so until the sixth edition of 1873). He did, however, use the word "evolved." It appears once, as the very last word in the book, the final word of a justifiably famous paragraph.

> There is grandeur in this view of life, with its several powers, having been originally breathed by the Creator into a few forms or into one; and that, whilst this planet has gone cycling on according to the fixed law of gravity, from so simple a beginning endless forms most beautiful and most wonderful have been, and are being evolved.

It is important to remember that Darwin was no Darwinist. He could hardly have used the words "evolution" or "evolved" in the sense we generally understand them today, given that it was his own work that was largely responsible for altering the balance of their usage—from Cicero's unrolling scroll, to the transformation of organisms over geological time. We, however, are in a different position. To us, the shade of Darwin looms large. His insights have colored the way we think of ourselves and our place in nature.

So, when Darwin used the word "evolved," it was in the earlier sense, of something unfolding. Creatures would appear, perhaps in successively more elaborate forms, from simple beginnings—perhaps as an analogy with the production of a shoot from a seed, or a frog tadpole from a mass of spawn. Darwin was a great believer in the power of analogy. After all, his entire argument about natural selection was based on just such a comparison with the "artificial" selection that stockbreeders use to enhance the desirable traits in their charges.

Darwin, therefore, used the word "evolved" to mean growth and development of a complex form from a simpler one, and used it to draw an analogy with the altogether grander process in which life itself would from simple beginnings become more diverse, elaborate, and complex. Darwin had a term for this process to which evolution was a mere analogy: he called it "descent with modification," a much less loaded term than "evolution."

In general, though, when scientists in Darwin's time and earlier referred to the gradual change of species—what we today call "evolution"—they used the word "transformation." If evolution meant the unfolding of individual organisms, from seed to shoot, from egg to adult, then transformation meant the change in form of entire species, usually (though not necessarily) from simpler forms to more complex ones.

The two processes—evolution and transformation—were analogous, but distinct. Today, though, they have become conflated. When most people today talk of "evolution," what they mean is "transformation." This conflation has had the consequence of conferring a sense of direction and choreography onto the idea of Darwinian evolution. This is why people, when they think of "evolution," imagine (for example) a series of individuals, each one an improvement on the one before, and if there are gaps in the series, they are "missing links"—pieces in a metaphorical chain whose beginning, end, and intermediate progress are already known.

There are deeper roots to this conflation, however, but before I get to that, I must tempt you into a little digression about the nature of and evidence for Darwin's descent with modification.

Earlier I mentioned that the "community of descent" provides much evidence for descent with modification. By this I mean that all forms of life are organized in fundamentally the same way, down to the minutest detail, supporting the view that all life shares a common heritage.

It's worth considering this in a little more detail. As far as we know, all organisms owe their structure to the peculiar chemistry of the element carbon. Carbon atoms readily bind with one another and with atoms of other elements (notably oxygen, hydrogen, nitrogen, phosphorus, and sulfur) to produce highly elaborate molecules, sometimes disposed in long chains of smaller, similar units strung together. So it is that all organisms so far discovered carry genetic information in the form of long, carbon-based, chain-like molecules called nucleic acids, either DNA (deoxyribose nucleic acid) or the related form RNA (ribose nucleic acid). This information specifies the structure of a different set of chain-like, carbon-based molecules called proteins, and does so using a code that's the same (albeit with minor variations), irrespective of the organism concerned. All organisms more complicated than viruses have cells, bounded by a membrane constructed of two layers of carbon-based, chain-like molecules called lipids, sometimes bound in an extracellular matrix made of chain-like carbon-based molecules such as cellulose or collagen. The contents of the cells are pretty much the same, irrespective of the organism in which they occur.

The similarities between creatures at this most detailed level are so great that it's a wonder that organisms as a whole come to look so different—from the worms burrowing beneath Darwin's tangled bank to the birds and insects flitting above it. This underlying sameness is such compelling evidence for descent with modification that it would, according to Richard Dawkins (in his book *The Greatest Show on Earth*), stand alone, even had no fossils ever been discovered.

Why is the evidence so strong? Because life needn't have been arranged like that. It is possible to imagine systems that have some of the properties of life that use only some of the above features, or none. It is also possible to imagine a situation in which different living organisms sharing the same planet have fundamentally different constitutions. The fact that all life, no matter how various in form, is specified so minutely according to the same recipe suggests that all living creatures descend, ultimately, from a creature that had all these same fundamental features of inheritance and construction.

So much for descent: what of modification? Darwin supposed that the pattern of inheritance might vary, the offspring of parents becoming sorted by natural selection, so that the offspring would come to look different from their parents. These differences would accumulate, and the offspring would spread and diversify. As with offspring and parents, so, eventually, with new species arising from existing ones. It is a testament to Darwin's perspicacity that even though Darwin had no clue about the mechanisms of genetic variation, his suppositions have been borne out, innumerable times, and in exquisite detail.

Darwin imagined that life, governed by such a process, would be connected in a treelike pattern, rather like a family tree, with one ancestor at the bottom—the root and trunk—and progressively more (and more diverse) descendants as the branches and twigs. Darwin's conception of the treelike pattern of evolution formed the only illustration in the *Origin*. Darwin's innovation was his invocation of a process, natural selection, acting in the here and now, which, when summed over history, produced this pattern.

In geometrical terms, a tree is a box of boxes, a set of sets: one trunk gives off a number of branches, each of which gives off a bunch of twigs, each of which bears several leaves, and so on. The idea that life can be catalogued as a system of nested sets goes back to Aristotle, but it was formalized in the eighteenth century by the Swedish botanist Linnaeus, who originally devised the hierarchical means of classification we use today, in which species (*erectus*, *sapiens*) are grouped into more inclusive genera (*Homo*), which in turn are grouped with other genera into orders (Primates) and with other orders into classes (Mammalia). Linnaeus's conception of life was profoundly and inevitably pre-evolutionary: he was organizing life simply as he (and everyone else) saw it.

Scholars before Darwin thus had two distinct phenomena to explain. First was evolution—sometimes called generation—in which a small and simple germ was elaborated ("evolved") into a large and complex adult. The second was the apparent arrangement of life in a hierarchical or treelike fashion.

The analogy Darwin drew between evolution and transformation was not his own invention. Editions of William Harvey's *Exercitationes de generatione animalium* (1651), one of the earliest works in the modern era to address the question of generation, bore engravings illustrating Zeus holding an egg from which all manner of creatures poured forth, with the legend "Ex Ovo, Omnia"—everything comes from the egg—a slogan that could be applied to generation and transformation with equal facility.

A more explicit connection between the two processes was drawn by the adherents of "nature philosophy," a tendency popular in the late eighteenth and early nineteenth centuries, and particularly associated with the poet, protoscientist, playwright, and all-around egghead Johann Wolfgang von Goethe. The nature philosophers were inclined to be somewhat romantic, which doesn't always go down well among scientists, and it's easy to make fun of them nowadays. However, they made two vital contributions to biological thought—one somewhat mystical, as one might expect; the other highly practical.

Although people saw life arranged as a tree, they also noticed that trees grow upward, from the ground; that you need a ladder to climb a tree if it is tall; and that it takes more effort getting to the upper branches than sitting on the ground. The treelike arrangement was therefore in accord with the ancient idea of the "great chain of being," in which living creatures occupied a station in life according to their structure, the simpler ones (worms, insects, and so on) toward the bottom, the more complex ones (fishes, birds, mammals) toward the top. Human beings would—noblesse oblige—occupy the topmost rung, above the apes, but below the angels.

At first, this arrangement was simply a statement of the order of creation. There was no sense in which creatures on a lower rung could be transformed into creatures on a higher one. Some thinkers, however, began to question why the tree should be ordered in the way it was, rather than in any other way, and began to imagine processes whereby creatures might be transformed.

Perhaps the most famous exponent of transformation before Darwin was Jean Baptiste de Lamarck, who outlined a scheme in his book *Philosophie zoologique* (1809) in which creatures would be driven to transform by an inner force or *besoin* (need) in response to their environmental circumstances, and such transformations would be inherited by any offspring. Thus the canonical picture of giraffes extending their necks ever longer to reach the highest leaves, and passing the results of their exertions onto baby giraffes, which would tend to have longer necks than their parents. This idea sounds quaint today, but Lamarckism was a theory with legs.

Today we are inclined to think that after the publication of the *Origin*, Darwin's ideas just went from strength to strength (such is our view of history as forever progressive), but this is not the case. Natural selection required that creatures provide a constant source of variation on which this selection could act. In Darwin's time, though, no such mechanism was known. The discovery of genetics around the turn of the twentieth century was to answer the question and so rehabilitate Darwin, but for half a century—between Darwin's death in 1882 and the reconciliation of evolution and genetics in the late 1930s—evolution by natural selection was in eclipse: influential scientists turned away from Darwinism for want of an explanation of variation, leaving evolution as not much more than a set of just-so stories. William Bateson—the scientist who would later coin the term "genetics"—was typically scathing.[8] "In these discussions [of evolution] we are continually stopped by such phrases as 'if such and such a variation then took place and was favourable,' or, 'we may easily suppose circumstances in which such and such a variation if it occurred might be beneficial,' and the like . . . 'If,' say we with much circumlocution, 'the course of Nature followed the lines we have suggested, then, in short, it did.'" As a result of this Darwinian vacuum, many mainstream thinkers continued to favor Lamarckism, so much so that it formed the grounding of university-level textbooks such as E. S. Russell's classic *Form and Function* (1916).

The nature philosophers looked at the pattern of life, but rather than Lamarckian *besoin*, a mechanism that was actually meant to cause transformation in the real world, they saw in each successively more elaborate form a more concrete manifestation of some ideal, cosmic striving toward perfection that would reach its acme in Man (with a capital *M*). Creatures in the real world were imperfect expressions of a transcendental ideal. No actual transformation was meant to have happened.

The practical aspect of nature philosophy came with nature philosophers' approach to the problem of generation. The problem of generation was working out how a seemingly unformed germ (such as a seed or egg) evolved into a complex, adult creature. Where did all that complexity come from?

Some scholars thought that it appeared out of nothing, whereas others, the so-called preformationists, thought that the adult form was there all the time, just in some occult, condensed form, waiting for the right cue to unravel. The problem was that investigating the subject directly proved impossible, and by the end of the eighteenth century the subject had reached an impasse. The problem couldn't be solved until the adoption of the cell theory, in the 1840s, and with that, the invention (one is tempted to say "evolution") of staining techniques whereby translucent, filmy cells could be made visible under a microscope. Only then was it realized that new organisms arise from the fusion of male and female sex cells (sperm and eggs) followed by a complex series of elaborations ("evolutions").

In the meantime, though, the nature philosophers took the view, possibly informed by their somewhat mystical outlook, that the earliest stages of generation might be forever hidden from view, impossible to discover even in principle. If this sounds familiar, it should—astrophysicists have adopted the same view about the birth and very earliest moments of the universe, ruled by physics beyond current theory to explain, and probably beyond any capacity of experiment or observation to penetrate. But that doesn't stop astrophysicists observing and theorizing about the history of the universe after that mystical instant of birth, and nature philosophers took the same view of generation. If the earliest moments of generation could not be seen, there was still a wealth of information to be gained about embryos, and how they grew and developed.

When German-speaking embryologists such as Karl Ernst von Baer and especially Ernst Haeckel, who had been drenched in the culture of nature philosophy, came to look at the embryology of various creatures, they found that the stages through which a developing organism "evolves" reflects its station in the grand ordering of Creation. Creatures start from single cells, much like blobs of protoplasm. They then form into balls of cells, similar to lowly algae or sponges, which fold into cup shapes, blind sacs with an opening at one end—much like simple polyps. They then elongate, coming to look like lowly worms, with yet further evolutions demarcating successively higher states. The necks of human embryos, for example, show rudiments of the gill slits that perforate the throats of fishes. They have tails, which are reabsorbed, and just before birth, some babies are quite furry. The elision, therefore, became obvious. The great tree of life, the great chain of being—whatever one wants to call it—maps the evolution of every individual creature as it develops. To put it another way, the evolution of any creature goes through a number of stages, the last one of which determines its place on the tree of life. The canonical summary of this idea is "ontogeny recapitulates phylogeny." This concept was meat and drink to the nature philosophers, who could now see the archetypal ideas of creatures on the grandest scales played out everywhere in the dramas of individual development. As one nature philosopher put it: "What is the animal kingdom other than an anatomized man, the macrocosm of the microcosm?"[9] It was the nature philosophers then, who, when they became embryologists, made the explicit connection between what might otherwise have been seen as two quite distinct processes—evolution and transformation. Partly for this reason, one can lay the blame for today's muddled thinking about evolution at the door of the nature philosophers and their inheritors, especially Haeckel.

The nature philosophers did not see the natural world in terms of actual transformation, only as the expression of cosmic or divine ideals. Haeckel, though, became a firm adherent of Darwin's evolution, doing much to popularize it. Haeckel missed the essential metaphor of Darwin's tangled bank, however, and saw natural selection instead as a kind of motor that would drive transformation from one preordained station on the ladder of life to the next. This is the view of natural selection—as another word for the cosmic urges of nature philosophers—that some scientists[10] found exceptionable toward the end of the nineteenth century, leading to Darwinism's eclipse, yet is the view that has become ingrained in the public mind whenever the word "evolution" is mentioned. It is this Haeckelian bastardization of natural selection that's responsible for the arrows in figure 2, the engine that drives evolution forward, from simplicity to complexity, in a series of Ciceronian maneuvers with a definite beginning and a culmination in Man—as far from the undirected, contingent, and moment-by-moment actions of natural selection on the tangled bank as might be imagined.

And if we think that this piebald view of evolution, as forever progressive and improving, striving ever toward the transcendent light, is something espoused only by misinformed journalists and newspaper readers who know no better, we must think again. When I was an undergraduate, back in the mists of time (okay, it was 1981), my zoology textbook was the very latest edition of *The Life of Vertebrates*, by the influential, immensely respected, and very sensible zoologist, the late John Zachary Young. Here is Young summarizing the evolution of mammals, the group of creatures to which we ourselves belong.

> We shall expect to find in the mammals even more devices for correcting the possible effects of external change than are found in other groups. Besides means for regulating such features as those mentioned above we shall find that the receptors are especially sensitive and the motor mechanisms able to produce remarkable adjustments of the environment to suit the organism, culminating in man with his

> astonishing perception of the "World" around him and his powers of altering the whole fabric of the surface of large parts of the earth to suit his needs.[11]

Yes, you read that correctly—Young really does use the phrase "culminating in man." And if that's in a modern undergraduate textbook, written by an acknowledged authority, it is little wonder that people more generally find it hard to grasp what evolution (in the sense of descent with modification) is all about.

We can't put all the blame at Haeckel's door, however. When the *Origin* first erupted (there is no other word) into the public consciousness, commentators were less worried about the niceties of natural selection, still less that Darwin could not explain the mechanism of inheritance on which his theory depended, but about the challenge that Darwin's ideas made to established social orthodoxy. In place of a static social order, a possibility of change—of liberation, progression, advancement, improvement. What we would now call a left-wing thinker such as Harriet Martineau (who knew Darwin personally) and particularly Herbert Spencer (who coined the phrase "survival of the fittest") co-opted Darwinian evolution in support of a general theory of social evolution that had all the hallmarks of the directed, progressive strivings that one would see turning up everywhere from manifest destiny and Marxism to fascism and advertising.

The *OED* defines sense 10 of "evolution" as "progression from simple to complex forms, conceived as a universal principle of development, either in the natural world or in human societies and cultures" and cites Martineau.

It was Spencer, not Haeckel, who championed evolution among what we might now call the "chattering classes," in opposition to the nobility and the established church, and who wrote, just before the *Origin* was published, that "those who cavalierly reject the Theory of Evolution, as not adequately supported by the facts, seem quite to forget that their own theory is supported by no facts at all." The battle lines were drawn between the agents of political progress, marching forward with evolution as a kind of justification for social improvement, and the established orthodoxy to which evolution was seen as a threat. One sees the same lines drawn to this day, especially in the United States. It's a pity that somewhere along the line, the exquisite beauty and infinite subtlety of natural selection as a mechanism has been lost, trampled into the dust by the simplistic slogans of those who'd use evolution as a device to further their own ends.

The accretion of all this social, political, and philosophical baggage over the past century and a half has tended to dull any appreciation of the disarming simplicity and beauty of natural selection as a mechanism. All other schemes of transformation current in Darwin's day required strange and mysterious ingredients, such as Lamarck's *besoin*, or cosmic strivings for betterment favored by the nature philosophers—none of which could be seen or touched, and whose existence had to be taken on trust. Natural selection required nothing that couldn't be seen, touched, and appreciated by anyone.

Natural selection is unique in another way, too, for unlike all other theories of transformation, it has no inherent direction. Darwin's contemporaries and antecedents looked at the tree of life and invented processes to "explain" it that were directional and improving. Darwin turned this idea on its head. He came up with a simple process in which no particular direction was implied, but whose result would be the treelike pattern we see. The tree is just natural selection summed over history.

Natural selection, therefore, does not demand what we from our human perspective think of as "improvement." To go further, natural selection cannot be seen as evolution's guiding hand. It has no personality, no memory, no foresight, and no end in view. To be sure, it's easy to see that natural selection, if left to operate for long enough, will create the branching patterns of the tree of life in much the way that Darwin suspected it did. However, there is nothing in natural selection that allows you to predict any *particular* pattern that it might generate. This marks a crucial distinction between natural selection and earlier ideas of transformation that presupposed a ladderlike scheme with *Homo sapiens* at the top. In natural selection, the pattern we see was not preordained, manifest, or inevitable in any way. Stephen Jay Gould expressed this idea very well in his book *Wonderful Life*—if we could rerun the tape of life, we shouldn't necessarily expect the same result every time.

I'd like to go much further than Gould did. In a famous scientific paper, Gould and his colleague Niles Eldredge proposed that evolution would not always proceed gradually, according to the "insensible gradations" proposed by Darwin, but might in some circumstances proceed very rapidly, and in other circumstances not move at all.[12] This was the "punctuated equilibrium" model of evolution, much debated ever since. But the arguments about evolution's speed—and these arguments have been fierce and acrimonious—all rest on the assumption that there is a narrative to be uncovered, a story that might be read from analysis of the fossil record.

However, any patterns that we see in the fossil record are reconstructed by us, after the fact. Because the fossil record is so fragmentary and imperfect (a point that Darwin grasped with his usual percipience), it is easy for us to read into it any narrative we like and assume that this narrative must be the right one. It is only natural for us to compose a story that suits our own prejudices of evolution (driven by natural selection) leading to ever greater refinement. This is, however, a profound misreading of Darwin's ideas and reflects a failure to understand the uniqueness of natural selection as a mechanism of transformation. With natural selection, no fate is ever inevitable, unless reinforced as such by hindsight.

The blob of protoplasm in Darwin's proverbial "warm little pond" could have evolved into anything—or nothing. The fact that evolution took the course it did was a result of natural selection acting on it and its descendants, moment by moment, according to the environmental circumstances prevalent at each given instant. Looking back at the course of evolution from our privileged height, we naturally assume that the only course of evolution possible was the one that led to ourselves.

This idea seems to have made insufficient impact among science communicators, members of the public, and even some scientists. In the world at large, many evolutionary transformations and adaptations are assumed to have been imbued with purpose. For example, feathers are seen as adaptations that allow birds to fly, as if flight were somehow the manifest destiny of birds. That this idea is wrong is shown by the evidence, which suggests that feathers evolved many millions of years before birds took to the air, among dinosaurs that patently would not have been able to fly. It is even possible that some dinosaurs, having evolved feathers, lost them again. This kind of backward-reasoning, in which adaptations are seen as having a purpose in some great transcendental game that lasts for millions of years, is also widely seen in schemes of human evolution that suppose, for example, that humans stood on two legs in order to free up the hands for making tools, to nurse babies, and so on.

This style of reasoning, in which evolution is assumed to have a purpose or a goal, is naturally accompanied by an assumption of progress, very much in the pre-Darwinian style. The assumption of progression is not only a misrepresentation of evolution, but ignores most of what is actually going on.

When we strip away the assumption that evolution is progressive, we find a different picture, both richer and stranger. Most of what seems to be going on in evolution is not the acquisition of new, improved ways of living, but their wholesale loss. This is quite at variance with the picture of evolution most people have, of a march of greater complexity and improvement—a picture that, as I hope is becoming clear, is sometimes misinformed. The concept of loss is explored in the next chapter.

3: *Losing It*

Evolution by natural selection, then, is not a noble or divine force that carries organisms on tracks of inevitable and inexorable improvement from the past to the future. Once we've roasted that old canard and served it up with orange sauce, we can begin to demolish as spurious the case for human exceptionalism.

But there's a catch—such progressive and inexorable improvement seems to have been precisely what has happened. Over the eons, living things really do seem to have become more complicated. Simple creatures consisting of single cells, such as bacteria, evolved into complicated creatures consisting of trillions of cells, such as human beings. If "improvement" can be equated with "complexity," then there seems to have been a general trend, throughout the history of life, for complexity to increase.

It is said that it takes just one ugly fact to destroy a beautiful hypothesis—so how fares my contention that natural selection is a consequence of several circumstances acting together only in the here and now, without having any end in view?

There are (at least) three answers to this. The first was very well put by Stephen Jay Gould in his book *Full House*. Yes, complexity has increased—but how could it not? If the earliest life was simple and microscopic, the only way was up. That aside, complexity seems to have been the concern of the rather small subset of creatures that includes ourselves. Even today, most creatures are simple and single-celled, and almost all of these are bacteria. Bacteria swarm on (and in, and around) every surface in uncounted profusion. Anyone who has eaten reheated cooked rice and come down with poisoning by *Bacillus cereus* might be astonished to know that the symptoms of poisoning are apparent only if there are more than 100,000 bacterial cells per gram of food.[1] This means that you can still swallow hordes of germs—cities, dynasties, empires of them—without even noticing, and suffer no ill effects whatsoever. Unbeknownst to our everyday selves, our skins crawl with bacteria, and bacteria in billions infest our guts.[2] Were every living creature counted as an equal, the total sum of nonbacterial living creation would be utterly insignificant. Complex organisms, rather than representing a general trend toward improvement, seem to have been a somewhat esoteric diversion.

Second, it all depends on what you mean by "complexity." How can such a thing be measured, and can it really be equated with "improvement" in any simple way? The simplicity of bacteria is more apparent than real. Bacterial cells might look simple—they are usually spherical or sausage-shaped, and their innards seem entirely featureless—but they are supremely adaptable. Many have digestions far more robust than the most adventurous gourmand, and can live in conditions that would kill any human being (and virtually anything else) instantly.[3] Bacteria live in the upper atmosphere, and deep underground.[4] There are bacteria that live in dumps of toxic effluent and in radioactive waste.[5] There are even bacteria so tough that they can survive exposure to the hard vacuum and intense radiation of space.[6] My point is that there are other ways of measuring complexity than numbers of cells, or the numbers of different types of cells in any given creature, or elaborateness of construction—in other words, according to the criterion by which we measure all things, that is, ourselves. More than 150 years after *The Origin of Species* was published, we are still wedded to the cosmic urgings of the nature philosophers, and accept it as axiomatic that Man is the microcosm that measures the macrocosm. In terms of chemical complexity, however, bacteria are far more complex than Man.

The third answer is more involved than either of these two, and goes deep into the mechanics of complexity increase.

The evolution of complexity is a hot topic in modern biology. The late John Maynard Smith, one of the finest biological minds of the past century, broke down complexity into a number of discrete steps, each of which had to be overcome before complexity could increase any further.[7] These steps included (among many other things) the evolution of very simple bacterial cells into the complex cells with which we are familiar, with discrete nuclei and subcellular compartments. Science needs its visionaries, and few were more visionary than Lynn Margulis,[8] who was the first to elaborate the idea that complex cells developed from simple cells working together to such an extent that they merged to become a single organism.[9] This idea, once dismissed as far-out, is now very well established and can be seen in various stages of completion, even today.

In many situations, bacteria of different kinds work together in sheets or mats called "biofilms."[10] The first large organisms—reefs and mounds of mineralized bacterial biofilms called stromatolites—are built of colonies of different bacteria working together.[11] Before the evolution of animals that could graze on them, stromatolites were common (they still live in isolated places where the water is too rich in salt or other minerals for other creatures to tolerate), and bacterial biofilms coated the ocean floor.[12] Biofilms are still with us, thriving in, among other places, the lungs of people with cystic fibrosis, where they contribute to the deadly pathology of that disease.[13]

Beyond biofilms, though, there is much evidence that complex cells, such as those that make up our own bodies, were originally formed from associations of several different kinds of bacteria that became so commingled that they could no longer function independently. The mitochondria—small sausage-shaped bodies in all cells—are relatively closely related to a group of bacteria called proteobacteria.[14] They even retain a vestige of their own DNA. The chloroplasts—the green bodies that give plant cells their green color (which also have their own DNA)—are distant relatives of the free-living, light-harvesting blue-green bacteria that contribute to stromatolites.[15] The DNA complements of mitochondria and chloroplasts, though, are mere scraps compared with those of their free-living relatives, as most mitochondrial and chloroplast functions have devolved to the nucleus, in which almost all the DNA of cells is archived.[16] Mitochondria and chloroplasts cannot function as free-living entities. By the same token, the nucleus—possibly the vestige of another kind of bacterium—depends on bodies such as mitochondria for its energy needs. This kind of union, known as "endosymbiosis," is now known to have happened many times in evolution. There are some algae whose cells bear witness to not just one but two separate, independent symbiotic events,[17] as if these cells were Russian dolls.

Complexity exists, and complex cells evolved from simpler ones. My thesis that evolution shows no definite trend in the direction of improvement would appear to have run into a sticky patch. *Au contraire,* say I.

I shall explain.

If evolution by natural selection can be said to have any "point" at all, it is that a creature should do all it can to improve the chances of its own offspring living long enough to reproduce. Why, then, would a simple cell, working perfectly well on its own, subsume its life and many of its functions in a larger collective, in whose stake it would have at best a slice of the action, rather than the whole cake?

The reason, I think, is all about energy, economics, and risk. Reproduction entirely on one's own terms is an expensive and exhausting business, and the expenditure might not always pay off. Economies of scale apply as much to living organisms as to human industry. The net benefits of working together might outweigh those of continuing as an individual, and these benefits, such as gains in overall efficiency, might include surplus resources that allow greater specialization among the members of a collective, which in turn improves energetic efficiency still further.

These rewards might also include the ability, perhaps, to colonize new ecological niches that might be inaccessible to one's competitors, and to do so speedily and efficiently; and, crucially, therefore, the capacity to perpetuate one's genetic heritage far more effectively than one might manage if working alone.

To take just one example: the first plants to colonize the land more than 400 million years ago were small, encrusting things. But the competition for soil nutrients and light was so intense that plants soon formed associations with soil fungi called mycorrhizae to help them get the best out of the earth. The mycorrhizae, living around the roots of a plant, would extend that plant's network into the soil, helping it extract water and vital nutrients. In return, the plants would feed the mycorrhizae the sugars created during photosynthesis. Plants with mycorrhizae would grow better than plants without, colonizing more and different habitats and increasing opportunities for their offspring to grow—and for their attendant mycorrhizae to prosper. In turn, the mycorrhizae would enable the plants to grow in soils in which they might otherwise wither. Today, land plants and mycorrhizae are totally dependent on one another.[18] Meanwhile, the plants themselves soon evolved specialized cells that created hard tissues capable of supporting stalks and trunks that could grow upward quickly. Within a geological eyeblink, forests of tall trees sprang up, each tree trying to outdo the other for a share of the sunshine. And so below, with the mycorrhizae around the trees' roots forming a wood-wide web of underground nutrient transport.[19] Plants and mycorrhizae have achieved far more by working together than either could have managed alone.

The associations between bacteria to form biofilms and then cells; the association of cells to form organisms, in which the cells can then specialize; the further association of different organisms into systems of mutual benefit, as with plants and mycorrhizae, and even into entire interdependent ecosystems as Darwin described so eloquently with his picture of the tangled bank—on the surface, these can all be seen as step changes in complexity. Complexity just seems to ratchet up and up, so it's no wonder that people tend to see evolution as a ladder that can be climbed, or as a chain in which there might be "missing links."

Except, of course, that it's not as simple as that.

As I alluded to above, organisms benefit from forming associations with other organisms, but that benefit comes at a cost. When organisms associate, it is because the benefits of living in a group outweigh those of living alone. The cost, though, is the sacrifice of immediate control of one's fate. The only thing that matters in the calculus of evolution is that the benefits outweigh the costs, however marginally—which means that if the benefits are large, then the costs will be only fractionally less.

Consider, if you will, a mitochondrion in a cell. Eons ago it was a free-living bacterium with its own DNA and could reproduce entirely on its own terms. Now, though, almost all its DNA has migrated to the nucleus, which regulates almost every aspect of its life. The mitochondrion cannot function on its own, and has been reduced to an energy factory, producing power not just for itself but for the whole cell. To be sure, the mitochondrion benefits from the fact that much of the work of its own maintenance has been contracted out to other parts of the cell, but the cost for this convenience is its former autonomy—and with that, its complexity.

The increase in complexity of the whole, therefore, is paid for by the complexity of the individual parts, and, in terms of the numbers of individual organisms subsumed into the greater whole, the total amount of complexity might be said to have *decreased*. If the net benefits to all organisms of living in an association have increased to allow specialization of its members, it follows that complexity is traded for efficiency. After all, a simple cell in which the mitochondria, chloroplasts, and nucleus are still, more or less, separate organisms will be both more complex and less efficient than a single organism with a pooled resource of DNA and a division of labor among much simpler components.

My contention therefore is that the seeming rise in complexity hides a deeper truth—evolution is not just about gain, but about loss. Once one gets away from the idea that evolution does not, in its own nature, demand an increase in complexity, one can see that any apparent increase in overall complexity is driven by a loss of complexity among the individual components that make up the whole.

This makes sense once you think of natural selection not as a driver of improvement as a matter of destiny, but the sum of all those circumstances that keep a creature alive only according to its present needs. Natural selection will ensure that organisms will do just enough—and no more—to exploit an advantage, however minuscule, for their progeny. If this means that they will lose a great deal of complexity in return for the marginally improved likelihood of passing on their genes that symbiosis or association might offer, then they will make that trade.

The fact is that bodies are expensive to build and maintain, and any creature that can get someone or something else to do the work instead will have the edge on a creature that insists on doing everything itself. There is a selective advantage, therefore, in being as simple as possible.

When people think of examples of the perfection of evolution (or, as it may be, the designs of the Creator), they tend to think of the evolution of beautiful structures such as the human eye or the tail of the peacock. Appearing very much further down the bill are parasites, creatures whose existence derives from the exploitation of other creatures, sometimes with grotesque, painful, and even lethal consequences.

Parasitism as a habit is hardly unusual. If you have ever dissected or gutted a wild animal—not a creature carefully bred for sport, science, or the table—you would no doubt have been amazed by the sheer load of parasites carried by an otherwise quite normal wild animal.[20] I remember as a schoolboy slitting open a freshly caught dogfish, and finding that its insides mostly consisted of worms. The animal was so full of worms that the poor fish resembled nothing so much as a sports holdall filled with wet spaghetti.

As a consequence of living off the efforts of others, parasites often become much less complex in form than their free-living relatives. Examples abound: one of my favorites is a creature called *Sacculina* that parasitizes crabs.[21] The mature adult is no more than a featureless blob, living on the crab and sending rootlets throughout the hapless host to extract its juices while it still lives. If you had to guess at the affinities of *Sacculina*, you'd probably say it was a fungus, but the truth is far more surprising. The true nature of *Sacculina* is betrayed only by its free-living larval stage, showing that it is, in fact, a kind of barnacle, but after the larva finds a crab to infect, it loses its shell and limbs, and indeed any obvious trace of its heritage, and becomes devoted to living off its host.

Sacculina might lose its shape, but it still consists of cells and tissues, and has its own complement of digestive enzymes and so on, all the better to consume its host. It has contracted out the services of locomotion, feeding, and much else to the crab, and for this gain it has traded its own limbs, mouthparts, sense organs, shell—just about every trace of its own crustacean heritage.

But parasitism can go a lot further than that.

Mycobacterium tuberculosis, the bacterium that causes tuberculosis, is a close relative of the leprosy bacillus, *Mycobacterium leprae*. But compared with the tubercle bacillus, the agent of leprosy has lost most of its genes. The tuberculosis bacterium has around 4,000 genes, compared with the 2,700 or so of the leprosy bacterium—of which at least 1,100 are known to be nonfunctional.[22] With little capacity to provide very much for itself, the bacterium relies on its human host for the means to go on living. It is, in fact, so feeble that it can hardly manage to reproduce on its own. Given that drugs against bacteria work best when bacteria are reproducing, this explains why this bacterium, weak though it is, is very hard to kill. Far from being a matter of survival of the fittest, the evolution of leprosy shows that there are advantages in weakness. The race does not always go to the strong. As a parasite, the leprosy bacillus has gone much further than *Sacculina*, which still, at least, maintains its own metabolism. But by contracting many metabolic services out to its host, and shedding many of its genes, *M. leprae* has arguably become less complex than its relative *M. tuberculosis*—and has become a more perfect parasite.

But not as perfect as it might be.

If the leprosy bacillus is alive, if sickly, some even smaller parasites can be described as hardly living at all. These are the viruses.[23] In general, viruses stand to bacteria as walnuts to watermelons. These creatures (I use the term loosely) are reduced to a few genes packaged into a protein coat. They have no digestive enzymes, no prospect of acquiring nutrients or digesting them, and no means of reproduction. They are in fact completely inert unless they can infect a cell (whether a bacterium or something more complex), whereupon they hijack the host's own biochemical machinery to produce more viruses. Viruses, then, look like the perfect parasites. They have lost just about everything except the inviolable essence of their existence—their genetic material—and use the services of other creatures to duplicate that material and spread it around. Given that viruses can't exist without more complex cells to parasitize, it is likely that they evolved from more complex organisms, refining and honing and streamlining themselves until they had lost all but the essentials. What might these organisms have been?

Most viruses have only a few genes—less than half a dozen—but there are some large and peculiar viruses, the so-called mimiviruses, which have more than 1,000 genes, making them as complex, genetically, as some bacteria.[24] This suggests that at least some viruses are stripped-down bacteria. However, it could be that other viruses are rogue genetic elements that have broken away from more from complex creatures.

It's hard to imagine parasites more reduced—more perfect—than viruses. But they exist. Amazingly, mimiviruses can be infected by tiny viruses, known as virophages,[25] and there are other viruses, the so-called satellite viruses, that cannot infect a cell unless riding shotgun with a more capable, larger virus.

And yet there are parasites more perfect still. As if to demonstrate the point that complexity is made possible by the simplification of its components, the ultimate parasites are part of us.[26]

Many genes in our own genomes once came from viruses that have completely lost the ability to create their own protein coats, and can reproduce only by inserting themselves into our own genomes. These creatures—entities—are called LINEs (short for long interspersed elements) and have sacrificed almost every shred of their separate identities. They were once retroviruses, that is, viruses whose genomes are made of RNA that is "reverse transcribed" into the DNA of the host. They contain just two genes—one for an enzyme called reverse transcriptase that effects this process, and another called endonuclease that cuts the host DNA, enabling the parasitic DNA copy to slip in. Although they can, in theory, jump around the genome like this, almost all LINEs known have long since lost this ability: they can only reproduce when the genome of the host does so. In effect, they have become part of the genome of the host. About a fifth of the DNA in the human genome consists of old LINEs strung together end to end, slowly mutating into randomness, like so many train carriages rusting, forgotten, in long-abandoned sidings.

But not even LINEs get the prize for being the ultimate parasite. That award goes to the so-called short interspersed elements, or SINEs, which are very short sections of DNA that lie in wait for a LINE endonuclease to make a nick in the host DNA to allow LINE insertion—and slip in ahead of it. LINEs in the genome are accompanied by retinues of SINEs in the way dogs have fleas, and SINEs make up around 11 percent of the human genome.

If LINEs have almost no genes, SINEs have none at all. All they have is a stretch of DNA (a sign of the SINE) that catches the attention of the host's enzymes, which transcribe it into RNA; this RNA is then reverse-transcribed by LINE reverse transcriptase back into DNA, which is then tucked neatly into place by the LINE's endonuclease. In this way, a LINE, a parasite with only two genes of its own, is parasitized by a SINE, which has no genes at all, but just the ultimate in self-reflexive identity, a genetic notice that says no more than "Pick Me! Pick Me!"

SINEs, therefore, are the perfect parasites. They are also the ultimate demonstration of my point—that as parasites devolve more and more of their own functions to their hosts, they lose more and more complexity, until there is virtually nothing left.

Now, you might regard as special pleading the idea that the complexity of a system can increase only at a cost of the complexity of its individual parts. You might likewise think of the example of parasitism, advanced in the cause of my argument, in like fashion—despite its ubiquity. However, the fact remains that evolution abounds with loss, and the more we discover about the evolution of various creatures, the more we see that loss has played a critical part in shaping the forms of life we see around us.

If symbiosis seems somewhat obscure, and the examples of parasitism I've chosen a little technical (you might never have seen a parasitic worm, or *Sacculina*; you've probably seen individual bacteria and viruses only in micrographs in books; and the existence of such arcana as SINEs and LINEs you must perforce take largely on trust), one can hardly argue with the concreteness of (say) birds. Birds are part of our daily lives. The smallest child knows what a bird is, and people who've heard of neither SINEs nor *Sacculina* can probably name many different bird species. Because of their ubiquity, beauty, and undeniable charm, the birds constitute perhaps the most intensively studied animal group. If the study of (say) parasitic crustaceans that eat the tongues of their fish hosts, only to replace them with their own bodies, is confined to a rather small group of specialists of epicurean taste, then the study of birds could hardly be more different, attracting flocks of professionals and veritable armies of knowledgeable amateurs.

One of the distinguishing features of modern birds is flight. Flight is an expensive pastime, such that the shapes of birds have been largely molded and subsumed to its cause—or so one might assume. The skeletons of birds are streamlined, with many bones fused together to form a rigid airframe. The bones are strong but hollow, making them very light. This hollowness extends to much of the rest of the insides of birds, too. The lungs of birds are connected to a system of air sacs that penetrates the entire body, even the insides of the hollow bones. As well as contributing to lightness, this air-sac system allows for a highly efficient system of gas exchange, as well as the cooling of internal organs heated by the fast metabolism that flight requires. Birds, like mammals, are warm-blooded, and run hot.

The outsides of birds are equally distinctive, being clothed in feathers. These remarkable appendages[27] permit the bright and varied coloration of birds—vital to their often complex social lives—as well as creating a smooth, drag-free external surface, vital for rapid movement through the air. In addition, many feathers are ideally shaped as airfoils, whether individually or acting together. The presence of feathers seems to be, quintessentially, the feature of birds that marks them out from any other creature.

That the shapes of birds seem to have been subsumed to the needs of flight is testament to the enormous energetic cost of this habit. Therefore it should be no surprise to you—having read this far—that birds conspire to lose it at every opportunity. All birds alive today are thought to have descended from flighted ancestors, so it is remarkable to see how many are flightless. The power of flight has been completely lost in two entire orders of bird—the ratites (ostriches and their relatives) and the penguins, and many other bird groups have representatives that are flightless. Birds that find themselves on remote islands free from land-living predators routinely give up flight as an expensive luxury. The Galápagos Islands have their flightless cormorant, and the extinct dodo of Mauritius was a gigantic, flightless pigeon. The kakapo of New Zealand is a large, flightless parrot. Some birds that retain the power of flight aren't actually very good at it: the world has yet to see chickens migrate.

Many flightless and other ground-nesting birds, particularly on remote islands, have been particularly vulnerable to extinction from the depredations of human colonists and their retinues of cats and rats. One thinks not just of the dodo but the moas of New Zealand and the gigantic elephant birds of Madagascar, and whole hosts of birds endemic to remote island groups such as Hawaii. But even before people and their domestic animals and pests turned up to spoil things, birds gave up flight wherever they could. Not long after the dinosaurs became extinct, the role of top land predator was taken by gigantic, carnivorous, flightless birds, the phorusrhacids, relatives of modern cranes and rails. Even further back in time, in the Cretaceous period, when dinosaurs were still running round and bumping into one another, the flightless seabird *Hesperornis* ducked and dived in the seaway that once bisected North America from north to south.

When I said, a few lines above, that the whole frame of birds seems to have been adapted to the habit of flight—well, that was the view until relatively recently. In the past twenty years or so it has become generally accepted that birds are the closest living relatives of dinosaurs, and it so happens that many of the features that we see in birds, and that were generally thought to have been unique to birds and specific adaptations for flight, also turn up in dinosaurs, many of which were large, heavy, and as aerodynamic as a sack of spanners. The hollow bones of birds, combined with the loss or fusion of many bones, especially in the limbs, are found in dinosaurs, even quite large ones, and I think I'll have done seen 'bout everything, when I see a *Brachiosaurus* fly. There is good evidence that the bodies of dinosaurs were full of air sacs connected to the lungs; that some of them folded their forearms, with the hands backward, just as birds fold their wings; that dinosaurs such as *Velociraptor* had wishbones, a distinctive feature otherwise only seen in birds; and that some dinosaurs incubated their eggs just as hens do.[28] But the most dramatic evidence among many features once thought distinctive of flying birds is the presence of feathers in many dinosaur species.[29]

Much research over the past few years has shown that the origin of birds lies somewhere among a group of dinosaurs called theropods, specifically small theropods collectively known as Paraves (near-birds).

This group includes *Velociraptor* and the fearsomely clawed *Deinonychus* as well as the remarkable "four-winged" gliding dinosaur *Microraptor* and the feathered hunter *Sinornithosaurus*.[30] It is among the Paraves (and their close relatives the oviraptorosaurs, such as *Oviraptor* and the enormous *Gigantoraptor*)[31] that one finds the greatest concentration of feathered dinosaurs. Somewhere in this group lies one of the most iconic fossil species ever discovered—*Archaeopteryx*. So iconic, that knowledge of its significance has permeated society at large.

My elder daughter, then age three, was a very frustrating kindergarten student. When all the other children were paying attention and behaving themselves, Gee Minor would whizz around the playground, arms outstretched, shouting "I'm *Archaeopteryx*! The first bird!" When Mrs. Gee or myself would come and collect her, we would suffer remonstration from her teacher. "Your daughter is *not* an *Archaeopteryx*," we'd be told: "she's a *little girl*." That the status of *Archaeopteryx* should be known and appreciated by small children should be a guide to its importance and the place it holds in the general consciousness of evolution.[32]

Archaeopteryx first came to light as a single fossil feather in 1861, soon followed by skeletons of entire animals, each with a halo of feathers, impressions on the very finely grained limestone in which these creatures had been entombed.[33] Only a handful of *Archaeopteryx* specimens have since been found, all from the same area of southern Germany. Apart from the feathers, arranged as beautifully on the wings as on any pigeon, *Archaeopteryx* looked very reptilian. Where modern birds have a short, stubby tail (the "parson's nose" of your Christmas or Thanksgiving roast), in life surmounted by a fan of feathers, the tail of *Archaeopteryx* was long and bony. Where modern birds have a toothless beak, *Archaeopteryx* had jaws full of teeth. *Archaeopteryx* lacked the large keeled breastbone for the attachment of flight muscles that is typical of modern birds. And so the list of differences goes on. But there were similarities, too—studies of the skull of *Archaeopteryx* show that its brain was similar to that of birds in many ways,[34] and it also had the hollow bones typical of birds today.

At the time of its discovery, and for a century or more after, *Archaeopteryx* was seen as a transitional fossil—a missing link—between reptiles and birds, a wonderful vindication of Darwin's ideas only two years after the publication of the *Origin*. It is no surprise, therefore, that *Archaeopteryx* gathered the soubriquet of the first bird, and that in it was seen a tendency—a trend—toward the progressive loss of reptilian features (teeth, long tail) and the gain of more birdlike ones (feathers, hollow bones, keeled breastbone) seen in modern birds.

In retrospect, the days of *Archaeopteryx* holding its place in the hearts of small children and paleontologists as the first bird were numbered with the first account of a feathered dinosaur, *Sinosauropteryx*, in 1998.[35] Many of the reptilian features of *Archaeopteryx* were dinosaurian ones—and, more remarkably, so were those of its features once thought typical of modern birds, such as feathers and hollow bones.

In fact, the latest research suggests that *Archaeopteryx* was not especially closely related to modern birds, being more closely related to dinosaurs such as *Velociraptor* and *Deinonychus*.[36] Small children in playgrounds the world over will now have to shout "I'm *Archaeopteryx*! Just another feathered dinosaur!"

But here's the killer. Even if *Archaeopteryx* was only a first cousin of birds, it was still a flyer. So was it a representative of a tendency among ground-living dinosaurs to get airborne—to strain, perhaps to yearn, in a suitably nature-philosophical manner, for the wide cerulean welkin?

Well, actually, no.

The latest research shows that the more feathery, flight-inclined members of the group to which *Velociraptor* belonged also tended to be the earliest and more primitive members of the group—creatures such as *Archaeopteryx* (and others such as *Microraptor* and the less familiar *Xiaotingia*). It looks very much as if this group of dinosaurs started off with flying and gliding animals that tended to lose this capacity, rather than improve on it.

Archaeopteryx was, therefore, not a stage in the acquisition of flight in birds, but in its loss among a related but different group of dinosaurs whose later members, while feathered, did not fly. *Archaeopteryx* was not a harbinger of things to come, but a one-way ticket to extinction. As far as we can tell at the time of writing, the closest relatives of birds among dinosaurs were small, very peculiar, and nonflying feathered dinosaurs called scansoriopterygids.[37]

The latest version of the story of *Archaeopteryx* (which, it has to be said, remains highly controversial) turns the original on its head, and shows that the tendency to lose the habit of flight runs deep into the dinosaurian roots of birds—and that the phenomenon of loss, more generally, pervades evolution even to the extent of knocking one of our most treasured missing links off its perch.

Once one realizes the extent to which loss has shaped evolution, one starts to see it everywhere.

Among the tetrapods—the group of four-legged animals that includes most creatures with which we are familiar (including, as it happens, birds and human beings)—we see the loss of some or all the limbs in many different lineages, such as snakes, several kinds of lizards and amphibians, and whales: four-legged animals, therefore, that have lost most or all of their legs. A recent discovery and the cause of much hilarity in the press corps is that human beings count among their unique attributes the loss of spines on the penis, a feature found more generally in other animals.[38] Loss is pervasive.

The take-home message of this chapter is that it is very hard, objectively, to decide which features of organisms are primitive and which advanced, especially if one is wedded to the view that the function of natural selection is to produce ever greater refinement and complexity. If a creature gains more mates, more resources, more short-term advantage by losing a structure rather than gaining one, then it will do so, and posterity can look after itself. The example of *Archaeopteryx* shows that the very idea that there can be such things as "missing links" represents a fundamental misunderstanding of how evolution works.

The term "missing link" should be expunged ruthlessly from our vocabulary. Journalists who use it should be subject to some embarrassing sanction, such as that in the probably apocryphal story that when the staff of the outgoing President Clinton left the White House to make way for the entourage of the incoming George W. Bush, they removed all the *W* keys from all the keyboards in the building.

This sanction should apply not just to the description of fossils, but of syndromes in modern humans in which patients appear to exhibit atavisms—throwbacks to some earlier stage of evolution. Much play, for example, was given to a family in Turkey whose members had a tendency to walk around on all fours.[39] Another better attested example concerns a gene called *FOXP2*, which seems rather different in modern humans compared with its form in other animals, and whose mutation is associated with a condition in which patients have great difficulty speaking and forming words.[40] Is *FOXP2* a "language" gene? Caution should be exercised in both cases. Such pathologies are the results of mutations in modern humans, and we have no way of knowing if they can tell us anything much about evolutionary history. In other words, they might be very revealing of how things are, but not how they got that way. Such examples perhaps say more about our own prejudice toward a "progressive" view of evolution.

The only direct evidence we have about the past comes from fossils. But fossils are mute. It is we who tell their stories for them, and these stories are likely to flatter our prejudices as much as reveal what is really there.

To make matters worse, fossils are so scarce it's a wonder we can use them to say very much at all—any patterns we're likely to learn from fossils are likely to be as provisional as our interpretations of the fossils themselves. This is the subject of the next chapter.

4: *The Beowulf Effect*

Charles Darwin was much exercised by what he called the "imperfection" of the fossil record, and viewed it as one of the chief difficulties of his theory. He was, perhaps, overdoing it—as I have discussed, subsequent research on the similarities between extant creatures, down to the molecular level, provides dramatic evidence for the community of all life. Darwin would have been safe had no fossils ever been discovered. It remains the case, however, that fossils provide direct evidence of evolutionary change in the past, and reveal how creatures have adopted many strange shapes not seen among organisms today. Without fossils, we'd be ignorant of *Archaeopteryx* and *Homo floresiensis*. The problem with fossils, though, is that no matter how strange they seem, we are overly inclined to see them as way stations in the canonical pattern of evolution we assume is there, the one that leads inexorably from primitive to advanced. It is all too easy to assume that *Archaeopteryx* is a "missing link" between reptiles and birds, and to dismiss *Homo floresiensis* as a genuine species because it doesn't fit in with deeply ingrained views about how the evolution of humanity "ought" to have happened.

Darwin was right, however, to have pointed out the imperfection of the fossil record. The fossil record is indeed imperfect, and in many ways more imperfect than we can imagine. In this chapter I shall show that it's so imperfect that one can never simply use what we've found to bolster preexisting notions of progress. More than that, the fossil record is so scanty that we cannot in all conscience ignore the lives and times of all those creatures that lived and died without leaving any trace of their existence. Such creatures probably constitute the vast majority of all creatures ever to have evolved. To ignore them would be as irresponsible as astrophysicists ignoring the majority of mass in the universe that appears to consist of "dark" matter, the nature of which is still unknown.

Finding fossils is a chancy business. We are therefore entitled to ask the following questions: What would our ideas about evolution have been like had none of the fossils we've discovered been found—but an entirely different collection unearthed instead? Would we still use this collection to justify a view of evolution based on progressive increases in complexity, culminating in Man?

I shall consider this by analogy with something we all know—the English language, so rich and strange, but something we take for granted. English is a first language for many, and a second for many more. English is among the top five most influential languages in the world,[1] and so ubiquitous that, as an English speaker, even the most remote parts of the world don't seem so far from home.

It didn't have to be that way. English is subtle and flexible, to be sure, but there is no particular reason, inherent in the language itself, why it should have achieved its present dominance over Latin, say, or Portuguese, or Malay, or, come to that, many of the other languages among the 6 or 7,000 spoken today.

The current success of English can be put down to historical accident, determined by two things. First, by its spread between the seventeenth and early twentieth centuries as the language of the British Empire, the most populous and geographically the most extensive commercial concern in history.[2] Second, by the fact that it just happened to be the language of those former British colonies that, as the United States of America, grew to eclipse their progenitor in influence and power.[3]

Neither the growth of the British Empire nor of the United States was inevitable. Had Wolfe failed in Quebec in 1759, say, and the Royal Navy lost to the French at the battle of Quiberon Bay in the same year,[4] I might be writing this book in French: or Spanish, perhaps, had the Armada not been blown off course in 1588. Or German, had Hitler pressed his advantage at Dunkirk. These examples might seem playful, but they are meant to be serious. Things that we take for granted, and assume to be the way they are through some inherent superiority or the inexorable machinations of destiny, might so easily have turned out differently.

Historians are now quite used to considering the might-have-beens as well as the documentary facts, and reconstructing "counterfactuals," scenarios of how things might have turned out had events gone slightly differently. These are more than merely speculative exercises. For example, documents still survive showing how the Nazis planned to govern the Soviet Union, had they managed to conquer it.[5] Pioneering Americans were brought up to believe in "manifest destiny," the doctrine that the United States would spread from coast to coast. That it did so might to some extent have been self-fulfilling prophecy. However, historian D. W. Meinig has challenged this, showing that the United States might easily have been much larger than it is—or much smaller—had certain policies been followed that were instead ignored, or put aside instead of being pursued.[6]

And so for the English language. Despite its modern currency, English began as a language—we call it Old English—that is as unintelligible to an untutored modern English speaker as, say, Swedish. It was spoken in England and lowland Scotland for about six centuries, from the invasions of Britain by the Angles and Saxons in the fifth century, until the Norman Conquest drove it to the brink of extinction. The few fragments of Old English literature that have come down to us from that remote yet immense period have survived thanks only to blind chance. For example, 30,000 lines of Old English poetry are known to us—all that's left of more than six hundred years of poetry and song. For comparison, Shakespeare's plays total some 150,000 lines, written over a period of twenty-four years. What's more, almost all Old English verse is found in *just four surviving manuscripts*, all written in the West Saxon dialect of Old English around the year 1000—which does not mean that we knew who originally composed them, nor in what language.[7]

Perhaps the best-known example of Old English that survives today is *Beowulf*. This is a long poem written in alliterative verse (a style characteristic of the period) concerning the adventures of the eponymous hero and his battles against a succession of monsters. The fact that *Beowulf* is a staple of the school and college curriculum, and can be found in the proverbial All Good Bookstores,[8] in the original Old English and in Modern English translation, inures us against the revelation that there is only one known manuscript of the poem—and that narrowly avoided being destroyed in a fire in 1731.

We are so used to the mass dissemination of information that it's hard for us to imagine a time before the invention of printing, when books were fabulously rare and expensive custom-made products, copied from an original (or from other copies), with great labor, and by hand. The fact that literature before the age of print could be reproduced only very slowly had an important consequence for knowledge. That is, it was once very much more fragile than it is now, much more prone to extinction. Given the prevalence today of print and electronic data storage, it would be very difficult, nowadays, to completely expunge all traces of *Hamlet*, say, or *Middlemarch*. Before printing, however, to put just one monastery library to the torch would have been to consign hundreds of unique manuscripts to total oblivion, irreparable and irretrievable.

If just one manuscript of *Beowulf* survives, one can hardly imagine the numbers of other works in Old English that once existed but that have been lost.

The facts of the manuscript speak for themselves. Apart from showing signs of fire damage, the *Beowulf* manuscript (you can see it on permanent display at the British Library in central London) is certainly a copy. It was made sometime in the eleventh century, presumably from another copy. The date of the composition of the original is not known—the poem might have been in existence for two or three hundred years before the single surviving copy was written. This suggests that the poem started as oral tradition; that there must have been a number of earlier written versions, all now lost; and that there might not have been a single, definitive, "official" version.

The single copy also shows signs of having been bowdlerized. The setting of the story is pagan, and concerns pagan values, but the copy we have was written many centuries after England had been Christianized. It is possible that the several references to Christianity in the poem are later additions, either in the manuscript we have—or in earlier versions, all now lost.

That tales once existed in Old English of which we now have no knowledge is illustrated by the use in *Beowulf* of words found nowhere else in the surviving corpus of medieval literature, but which are unlikely to have been neologisms created specially for the occasion; and obscure references to stories, whether of fact or fancy, that the contemporary audience would have found familiar, but which have since been lost and so mean nothing to us. Proof in the breach comes with an episode in *Beowulf* concerning a battle between two warlords, Finn and Hengest—an account of which same incident subsequently turned up in another fragmentary manuscript.[9]

What would our view of the past be like had no copies of *Beowulf* survived? And what of the alternatives? For example, if we think that the library of Old English is thin, of the native literature of England before the Anglo-Saxon invasion we know even less. What would our ideas of the languages, literature, and customs of the Dark Ages have been like had the single remaining manuscript of *Beowulf* been destroyed in that fire in 1731? What would our ideas have been like had we found instead an epic poem of King Arthur written in medieval Welsh? This is not idle fancy—the existence of such lost works was hinted at by Geoffrey of Monmouth in his *History of the Kings of Britain*, written in Latin in the twelfth century. Or what, perhaps, of tales in an otherwise obscure language such as Pictish, whose scant relics remain completely undecipherable?

The point of this is to show not only that history turns on a hair (the outcome of events is "contingent," as Stephen Jay Gould put it in *Wonderful Life*) but also that our present-day view of history is sensitively conditioned by those few and arbitrarily sampled fragments that have survived the ravages of time. I call this the "Beowulf effect."

As with fragile, unique handwritten scrolls from a thousand years ago, the chances of any living creature becoming a fossil are extremely remote. What's more, the fossils we have document an almost infinitesimally tiny, entirely arbitrary, and almost certainly unrepresentative selection of the range of living creatures that once existed, the preservation of any one depending very largely on luck. The fossil record shows the Beowulf effect in action.

The word "fossil" derives ultimately from the Latin verb *fodere*, which means "to dig." Baldly, fossils are things that are dug up. More specifically, fossils are physical signs of the presence of creatures that lived long ago, and that were buried. Fossils might constitute the actual physical remains of the creature—its bones, its shell, or even its DNA—but this is exceptional. More commonly, they are what happens when the tissues of a dead creature are replaced by minerals that percolate into the buried remains through the groundwater, creating a stony representation of the shape of the creature. The fossils of sea urchins that my family and I find on Cromer beach now and again aren't made of the actual material from which sea-urchin tests are made, but from chalk, or flint, a rock that forms when silica-laden groundwater percolates into chalk.

In some cases, especially when the creatures have become buried in an oxygen-poor environment such as the mud at the bottom of a stagnant lake, the bacteria responsible for breaking down the corpse will leave very detailed impressions of that corpse in the form of the deposits of their own mineral waste products. In other cases, fossils are the petrified impressions that a creature leaves in sand or mud—the cast of a shell or, more evocatively, signs of past activity, such as a trail of footprints, a bite mark, or a burrow.

What can't be emphasized strongly enough is that the chances of a creature leaving any trace at all in the geological record are vanishingly small. In the wild, many organisms—perhaps most of them—are eaten by predators. Should animals or plants live long enough to die without their bodies having been consumed by a predator, their remains are almost always recycled within days. Their soft tissues are soon eaten by scavengers, and any remnants are broken down to nothing by fungi and bacteria. The hard parts—whether bones or shells—are pulled apart and dispersed, and time eventually grinds them to powder. To stand any chance of fossilization—to become a recognizable memorial to an otherwise evanescent existence—the body of a creature must remain sufficiently intact until it becomes buried or otherwise put beyond reach of the normal agents of dispersal and decay. A fossil is therefore a sign of some rather unusual circumstances in which the normal course of events has been cheated.

Fossilization, if it occurs at all, almost always happens underwater, and to those parts of a creature that are most resistant to physical breakdown. This explains why the fossils we have are overwhelmingly those of the hard shells of animals that spent their lives in water, are therefore likely to die in it, and so stand a chance of becoming buried in the sediment at the bottom of the sea or in a lake. It is no coincidence that the collections of most amateur rock hounds contain fossils of sea creatures—clams, ammonites (the shells of creatures related to squid), belemnites (ditto), sea urchins, trilobites (marine creatures that looked rather like pill bugs), perhaps a fish or two, and the occasional bone of a marine reptile such as an ichthyosaur (whose bobbin-shaped vertebrae make excellent ash trays, I am told), but rarely the remains of ancient land animals such as dinosaurs. This is because land animals tend—of course—to die on land, and their remains disperse quickly before they can be buried. Fossils of land animals are usually what are left once the uneaten scraps of some hapless corpse get washed into a watercourse after everything else has finished with it. Transfer to water and subsequent burial break up the remains even further. This is why fossils of land animals are rarer than those of aquatic ones, and why even the best of those that survive long enough to be recognizable as the remains of living things are in general fragmented and in very poor condition. The majority of fossils of land vertebrates consist of teeth—this is because enamel is very much harder and more resilient than any other tissue, and is the last to be broken down.

Land animals that lived in dry conditions but close to water (a somewhat conflicting set of circumstances) are the least unlikely candidates for fossilization. Dinosaurs sometimes fall into this category, as do hominins. Creatures close to water sometimes fall in, or are pushed. There are, very occasionally, mass-death assemblages of dry-land creatures that have been overwhelmed by floodwater and quickly buried. The bodies of creatures that live in hot, damp tropical forests are almost always decomposed by other creatures and hardly ever fossilize; the bodies of animals that live at high altitude are broken up and decomposed long before they can be interred underwater in any recognizable state. Ancient hominins—at least, the ones we know about—lived in the lowlands, often near water (or in caves—another location that ups the odds of fossilization), so their fossils, while meager, are sufficient to mark their having existed. Chimpanzees are forest creatures, and although they have been evolving for precisely the same length of time as hominins, their fossil record consists, so far as is known, of just a few half-million-year-old teeth.[10] Gorillas, like chimps, live in tropical forests, sometimes at high altitude, and have been going their own evolutionary way for much longer than either chimps or hominins, but their fossil record is completely blank. Hundreds of thousands of generations of gorillas have come and gone, but apart from the creatures alive today or whose skeletons are preserved in museums, there is not one single sign, not even a scrap of half a tooth, to betray their lineage having existed, as it surely has done, for the past 7 or 8 million years.

Such is the process of fossilization: scrappy, chancy, biased, uncertain, and threadbare. There are, however, episodes of fossilization—remarkable because very rare indeed—in which living creatures are interred, often quite suddenly, to leave remains that are far better preserved than in the normal run, and which shed fractionally more than the usual murky half-light on vanished worlds.

The early birds and feathered dinosaurs of the Cretaceous of northeastern China, for example, were preserved in great abundance at the bottom of extensive lakes in association with volcanic ashfalls.[11] This has allowed preservation in such detail that the entire early history of birds and their relationship with dinosaurs has been completely revolutionized. Most of what we know about feathered dinosaurs comes from these deposits. These deposits are also rich in the fossils of early mammals. Isolated teeth or jawbones are usually all the traces that mammals leave as fossils, but the mammals from the Cretaceous of northeastern China are often preserved entire, complete with their furry coats.

The famous Burgess Shales of British Columbia, made more famous still by Stephen Jay Gould in *Wonderful Life*, preserve in exquisite detail an entire ecosystem of soft-bodied creatures from the Cambrian period, some 505 million years ago, creatures that just happened to have been suddenly buried in a submarine mudslide and preserved as shiny impressions on black shale. Sometimes the fossils are very hard to see unless immersed, when their gorgeous detail and strangeness emerges as if by magic.[12] Although fossils of marine creatures like those found in the Burgess Shales have since been found in strata in other parts of the world, some of them deposited since the end of the Cambrian itself, such fossils tell us far more about ancient marine life than can be revealed by regular garden-variety Cambrian fossils such as trilobites.

In another example, a freak sandstorm engulfed dinosaurs and other animals at a place called Ukhaa Tolgod in what is now Mongolia in the Late Cretaceous, some 70–80 million years ago,[13] burying them alive by a kind of three-dimensional instant photography—reminiscent of the preservation of the unfortunate citizens of Pompeii, overcome by the hot ashes and dust from the eruption of Vesuvius in AD 79.

There are several other instances of sudden, unusual kinds of preservation up and down the fossil record.[14] These cases are highly localized and possibly unrepresentative, but because they have the potential to yield so much more information on past worlds than fossilization does in general, their effect on our knowledge is disproportionately great. Had the freak mudslide that buried the Burgess Shales creatures not happened, how would our knowledge of ancient marine life have differed? What would we know had the mudslide entombed a completely different set of creatures? How would our knowledge of the relationship between birds and dinosaurs been affected had the conditions for preservation in those Chinese lakes been different, so that corpses that landed in them rotted more quickly, rather than less, or didn't preserve the feathers? Some rather smelly experiments on how bodies of fishes decompose have shown how our ideas of what ancient creatures looked like are very sensitive to their state of preservation.[15] On such tiny chances does the edifice of knowledge turn.

To return to my *Beowulf* analogy, consider these questions: How representative is *Beowulf* of Old English poetry? Did poets of that vanished age regularly write about manly heroes and horrible monsters, or was this exceptional? Did they, perhaps, tend more toward kitchen-sink or sitcom? Did they always write in alliterative verse, or did they occasionally stray into rhyming couplets? Indeed, can we say anything reliable about the totality of the Old English literary tradition, given the few examples that now survive? That we should be practical and do what we can with the evidence we have is no doubt the prudent answer—but we should never be lulled into thinking that any reconstruction we might build about the everyday repertoire of the scops and bards of yore is much more than educated fancy.

All the above presupposes that we have at least *some* Old English poetry—at least *some* fossils—to discuss, never mind how paltry the remains.

There are creatures in the modern world that are barely known as fossils. You'll remember how in the last chapter I made a big deal about parasites, and how their existence cocked a snook at the idea that evolution was necessarily a force for improvement and increased complexity. I mentioned that parasitism is very common, and that most creatures are infested with parasites. Many kinds of nematode worms (roundworms) are parasitic, living inside the tissues of most animals and plants. Nematodes also live freely in the soil, and even in rocks deep underground,[16] where they hunt for bacteria on which to graze. A student of nematode worms once remarked:

> In short, if all the matter in the universe except the nematodes were swept away, our world would still be dimly recognizable, and if, as disembodied spirits, we could then investigate it, we should find its mountains, hills, vales, rivers, lakes, and oceans represented by a film of nematodes. The location of towns would be decipherable, since for every massing of human beings there would be a corresponding massing of certain nematodes. Trees would still stand in ghostly rows representing our streets and highways. The location of the various plants

> and animals would still be decipherable, and, had we sufficient knowledge, in many cases even their species could be determined by an examination of their erstwhile nematode parasites.[17]

Nematodes are ubiquitous, and have probably been so for hundreds of millions of years, yet their fossil record is almost nonexistent. That "almost" is a big word, however—fossil nematodes have been found, preserved in amber, another unusual and chancy location for fossilization that yields spectacularly well-preserved fossils.[18] They have also been found in coprolites—fossilized feces—of dinosaurs.[19] But such occurrences only serve to underline my point. Nematodes are everywhere, and in everything, and (presumably) have been so for hundreds of millions of years. But their prehistory is betrayed only by a very few examples of fossils formed in rather peculiar circumstances.

Tapeworms, however, are another matter. They are completely unrelated to nematodes. All tapeworms are parasites, and are likely to have had a very long history and relationship with the animals (including humans)[20] they infest—but they have no fossil record at all. None.

Now, imagine that tapeworms existed for half a billion years, leaving no fossils at all, and became extinct. We would have no knowledge of their ever having existed at all.

From the above it makes sense that there must have been many kinds of creatures that once existed but that have vanished without trace. Until the Burgess Shales had been uncovered, we couldn't have known of creatures such as *Opabinia*, a swimming shrimp-like creature with five eyes on stalks and a flexible proboscis furnished with serrated jaws, a creature of a kind that nobody had even imagined existed.[21] It is possible that entire groups of unimaginably strange creatures have lived on this planet for millions of years but died out leaving no trace at all in the fossil record. If we had never found *Opabinia*, what other strange creatures might we have found instead? And how might the stories of life that we tell one another have been affected?

All this having been said, attempts have been made to quantify the degree of our ignorance, to assess the incompleteness of the fossil record as it applies to various groups of organisms.[22]

Completeness is relatively easy to assess on a small scale, though I use the word "relatively" with due caution. Let's say that you're digging in a quarry with the aim of finding a representative sample of all the kinds of fossil that might be present at that location. On the first day you find, say, five different kinds of fossil clam. The next day you might find two or three more, but after a while you just find more of the same kinds. As time goes on, the likelihood of your finding a kind of clam you hadn't seen before dwindles to almost zero (though never to zero itself). For all practical purposes, you could say that you'd excavated every kind of clam from that quarry that happened to be preserved there, as a fossil.

As we have seen, however, a number of factors influence the repertoire of the animals and plants that once existed in a locality that get preserved as fossils. Had the soil in which the dead creatures were buried been more or less acid, or more or less oxygenated; had the winds and currents been blowing this way or that; had the ambient temperature been somewhat higher or lower; had the groundwater been infiltrated by one mineral rather than another—all such things and more might have influenced the kinds of creatures more likely to have been entombed as fossils.

You might, in your quarry, have found every kind of clam that once existed there in the remote past, but try as you might—and completely without your knowledge—you'd never find a single momewrath, because momewraths would not have fossilized well in the conditions that entombed the clams so faithfully. You would have no way of knowing that in the ecosystem whose only vestiges are found in that quarry—that tangled bank—momewraths were the most abundant and dominant creatures, outnumbering even borogoves, both creatures being the prey of the utterly frumious bandersnatch. Clams were always something of a sideshow. But clams are all that's left, and momewraths, borogoves, and frumious bandersnatches have disappeared from the earth without leaving a trace. If this sounds fanciful, here is a real example. In the Doushantuo phosphorites of China you can find fossils of 600-million-year-old creatures preserved so beautifully that you can pick out individual cells. The fossils are all microscopic—anything larger than a pinprick is absent. Nobody really knows why.[23] Neither does anyone know what these creatures were.

You might contend that the case of the Doushantuo phosphorites seems like special pleading. They represent, it is true, a rather specific set of circumstances somewhat different from the usual run of fossilization. I invite you, therefore, to consider the conodonts. These are fossils of small but elaborately constructed arrangements of tooth-like elements of such abundance and variety that many rock strata are known by the species of conodonts they contain. The problem is, nobody had any idea about the kinds of creatures to which these toothlike fossils belonged.

Many different candidates were offered, more or less bizarre, but the case could not be settled because no fossils had been found that preserved conodonts and any associated animal in any convincing way.[24] One of the most peculiar candidates (in a pretty weird bunch) was a 320-million-year-old fossil called *Typhloesus wellsi*, found with conodonts in its insides.[25] Critics argued very reasonably that *Typhloesus wellsi* wasn't the conodont-bearing animal, but a predator that ate conodont-bearing animals, leaving only the conodonts to fossiliferous posterity. To this day, nobody knows what kind of animal *Typhloesus wellsi* was. All we had was a tiny glimpse of momewraths being eaten by bandersnatches of dubious frumiosity. Considering the conodonts as a whole, it was as if we human beings and all our works vanished utterly, all except for our dentures.

Eventually, fossils of soft-bodied, eel-like animals were found in which conodont elements were found arranged at one end, like teeth. A few more turned up just to show that this wasn't a fluke, and a consensus was reached that conodont animals were akin to fishes, although representing an entirely separate evolutionary experiment in aquatic vertebrate life.[26] It has to be said that not everyone agrees with this view,[27] but the fact remains that conodonts are so common as fossils that the oceans must, at one time, have seethed with these creatures—all now gone. All, that is, except for their enigmatic smiles, like so many million Cheshire cats.

Measuring completeness on a larger scale is even more problematic.

One of the most important repositories of paleontological information is the catalogue of fossil diversity first assembled by the late J. John "Jack" Sepkoski of the University of Chicago.[28] Sepkoski tracked down every report of every kind of marine invertebrate fossil ever found, charting their first and last occurrences in the geological record, and their ranges in time. Using Sepkoski's magnum opus, other paleontologists have sketched broad outlines of the history of life, noting—for example—epochs in which life seemed more or less abundant, in which entire "guilds" of creature replaced one another over geological time, and episodes of "mass extinction" in which life seemed almost to wink out altogether. Databases such as this have allowed paleontologists to approach the completeness of the fossil record in an altogether more scientific, quantitative way, applying statistics to the known unknown.[29]

Let's go back to that quarry where you've been collecting fossil clams. Imagine you are interested in the fossil record of just one species. You know that fossils of this species have been collected over a range of 20 million years, based on the first and last known occurrences and reports from perhaps a dozen localities in between (of which your quarry is one). Now, ask yourself this question: how "complete" is the fossil record of this species?

In one sense the answer is "not at all," given that you know of only a few fossils of this species, representing a spread of at least 20 million years. Perhaps billions of individual clams of this species lived and died during this period, in which its fossil record is infinitesimal. However, the record is just good enough to show that the species existed and survived for a span of time, so one can get a measure of whether this range of 20 million years bears any relationship to reality.

The first thing to appreciate is that the first occurrence of a fossil almost certainly does not represent the earliest existence of the species in life. Given that fossilization is rare, the species presumably existed for an unknown measure of time before one individual chanced to have been preserved. Neither does the last known occurrence of a fossil species necessarily record the last ever individual of that species before it became extinct.[30] How close can the fossil record get to reality, if fossilization is such an unlikely event? The answer lies in the density of the sampling between the two extremes. If, in that period of 20 million years, records of your species of interest are very sparse, then it is likely that the species tended not to fossilize well, which suggests that it lived long before its first record as a fossil, and long after its last. However, if a fossil species appears quite suddenly, is found pretty much everywhere in large quantities in closely spaced intervals of time, and then disappears without recurrence, we can be more confident that the time interval of 20 million years is a good reflection of reality. By the same token, one can predict that recurrence of the same species after a long gap would be unlikely.

Such recurrences do happen, however, and the reason is to do with geology and the circumstances of fossilization. Let's say that your clam, in life, preferred to live in shallow seas. Its extinction after 20 million years could be real—or it could simply reflect the fact that geological deposits representing shallow-marine habitats became rare, being replaced by deposits indicative of dry land. The discovery, perhaps, of a younger stratum representing shallow seas would be accompanied by more fossils of your favorite clam. Species that appear to become extinct but miraculously come back to life later on are known as "Lazarus taxa" after the biblical character whom Jesus raised from the dead (John 11:1–44). This phenomenon tells us something very important about fossils, and builds into my entire argument about the problems of building a narrative based on fossil evidence. That is, to become a fossil, a creature has to be found in the right kind of rocks. Rocks and rock types vary over time as much as the creatures whose remains are buried within them.

Recent work has shown that our measures of past diversity are quite sensitively affected by the amount of rock available in which fossils might be found.[31] The problem is that rocks (and the fossils they might contain) do not simply accumulate over time. An unknowable (and unknowably large) quantity of rock, created during the earth's long history, has itself disappeared—eroded, transformed, or sucked down into the ocean floor in the process of continental drift—taking its load of fossils with it to oblivion.

This sounds rather obvious, in hindsight. After all, you can't go looking for fossils in rocks that don't exist. However, it makes attempts to reconstruct past life—and account for its variation—rather tricky. It means that any trends we see in the history of life as reconstructed from the entries and exits of fossils might say very little about the history of life, but much more of the history of the rocks in which fossils are found. This implies that a great deal of life's story happened offstage, without report—and that we might be completely unaware of entire groups of creatures that once existed but have disappeared without trace. It could even mean that some creatures have undergone a kind of double extinction—that even after all representatives died, the few that remained as fossils were also expunged as the rocks in which they were entombed also perished. The stories we tell ourselves—of the rise of amphibians from fishes; of the domination of the earth by dinosaurs; of the subsequent rise of mammals, culminating in the proverbially zenithal apotheosis that is our own species—might very well be a sideshow, a tale that would not be supported were we made aware of the totality of all life that once existed on this planet.

I'll end this chapter with a discussion on how very close even the known fossil record is to being unknown—how close many species are to the lone copy of *Beowulf* that we can study and treasure. Sepkoski's compendium of fossils was based on marine invertebrates for the reason I have discussed above—that marine invertebrates have the best chance of all organisms of becoming fossils. The fossil record of animals that lived on land is much sparser.

Many species of dinosaur, for example, are known from just one or two specimens—if these specimens had not been found, the existence of that dinosaur would not have been reported. In such cases, chance effects such as rock type, and even whether a paleontologist happens to be there as a fossil erodes out of a cliff, before it is destroyed, have large effects—as large as the fact of the rescue of the single manuscript of *Beowulf* from that fire in 1731.

My favorite case of the Beowulf effect in action concerns a fossil of which you probably haven't heard. It is called *Palaeospondylus*, and on its own it's not much to look at—a tiny fish, between five and sixty millimeters in length. Quite a few specimens of *Palaeospondylus* are known, but almost all come from a single quarry at a place called Achanarras in northeast Scotland, in Devonian rocks that are around 380 million years old.[32] What can one do in such a situation? Estimating the geological range of a species is impossible if all one has is a single point, just one datum, so no one knows when *Palaeospondylus* first appeared, or when it went extinct. Paleontologists have debated the nature and identity of *Palaeospondylus* ever since its discovery in 1890, and have yet to reach agreement more than a hundred years later.[33] There have been suggestions that it was a larval form that would have grown up into one of the many other, larger fishes known from that part of Scotland of the same age. This idea makes a kind of ecological sense, given that so many specimens of *Palaeospondylus* are found together in a single place. It might be a snapshot in time of some kind of nursery, a nest or pond in which adult fishes sequestered their brood. There are problems, however—the fossils look far too bony to be larvae, at least of any fish we now know. But if the fishes are adults, we are left with a species without descent or antecedent, a species lost in time. Many, many species in the fossil record are like *Palaeospondylus*—known from just one locality, in which, perhaps, conditions for preservation just happened to be exactly right; in which the rock was not itself ground into powder with the fossils it contained; which just happened to have been unearthed by geologists and paleontologists who knew what they were looking at: fossils that had just that one, slim chance of making it into the realms of the known. We have no way of knowing the toll of species that were not so favored.

It's now time to apply this new, pragmatic, if rather chilly, view of the relics of evolution to the fossils that tell of the evolution of ourselves.

5: *Shadows of the Past*

With a working knowledge of evolution in our pocket, together with an appreciation that any trends or strivings toward perfection we see in evolution or the fossil record are readings we humans have made, after the fact, we're now ready to delve into the problem of human exceptionalism—the tendency to see ourselves as special products of creation, the result of an inevitable and predictable trend toward improvement and complexity.

In this chapter I'll pay particular attention to the extreme scarcity of hominins in the fossil record. Despite this scarcity, scientists still apply models of human evolution that are progressive and directed. Each new discovery of a fossil hominin is greeted by the press as a "missing link," when closer inspection shows that newfound fossils challenge our preconceptions at every turn. When this happens, scientists sometimes fight a rearguard action—the new fossil can't be new and different, but is really something known in another guise, such as a deformed human or an ape. We've seen this tendency at work in the discussion of *Homo floresiensis*, and it's not a new phenomenon.

Whatever happened in the past, everyone agrees that there are lots of human beings *now*. Not long before I started to draft this chapter, the world welcomed its 7 billionth human being. Although nobody could agree precisely which new baby was the 7 billionth, everyone agrees that 7 billion is an awful lot of people, and governments are beginning to wonder how many people the earth can realistically support.[1] Wherever you look, the world seems awash with people and the signs of their activities. This is of course entirely obvious in the world's teeming cities, but it is evident in the countryside, too, and even in apparently pristine wilderness. Much of the earth's surface has been changed to accommodate human needs for food and water. Human activity has started to change the earth's climate, affecting remote regions such as Antarctica where relatively few humans have trod. Beachcombers on the earth's most remote islands, uninhabited and far from the usual shipping lanes, find appreciable quantities of human refuse.[2]

When you look more closely, we human beings are a varied lot. Our most obvious feature—our skin tone—varies from deeply pigmented to virtually colorless, but humans vary in many other ways, both obvious and subtle, ranging from details of our anatomy to a whole host of differences in our body chemistries. We are not, however, as varied as we sometimes like to think. Compared with many other species, the genetic variation within *Homo sapiens*—the single species to which we all belong—is rather small. It is smaller, for example, than the genetic variation between the several isolated groups of chimpanzees scattered through central and west Africa—despite the fact that there are 7 billion of us and only a few hundred of them.[3]

It is easy to cast chimpanzees in the role of Our Ancestors. It is, however, only that, a role. Chimpanzees have been evolving away from our common ancestor for precisely as long as we have. However, chimpanzee variation does give us an insight into what human genetic variation might have been like for most of our evolutionary history. Humans might have been much scarcer—and much more varied.

The earth's current burden of humanity is an anomaly, for population has surged only relatively recently. When I was a boy, in the 1960s, there were only half as many humans as there are now. Before the invention of agriculture 10 to 12,000 years ago, there were probably no more than a million people on the planet at any one time.[4] This meant that population densities were very low, on average about one person for every fifty-seven square miles.

For most of the past few million years, humans and other hominins lived, like chimpanzees, as small, scattered groups, meeting one another only rarely. This meant that genetic variation between different groups was probably higher than it is today, tempered by the occasional exchange of mates. In most primates, it is usual for males to stay in the group in which they were born and raised, and for females to join other groups. This is true for humans, too—and is believed to have been true for early hominins such as *Australopithecus*.[5] On the whole, though, genetic variation in fossil hominins was probably higher than it is in modern humans, even though there might have been fewer individuals.

Rare species living in small groups are prone to becoming extinct by accident. The human genome shows that many human populations, especially outside Africa, were founded by small populations of individuals. It is this effect that probably accounts for humanity's rather low degree of genetic variation today.[6]

Several things follow from these arguments. The origin of new species requires genetic isolation between groups that might otherwise interbreed. Therefore, if groups were scattered and genetic variation high, it is likely that some of these groups diverged from one another to the extent that they would be considered as different species, at least when compared with the differences one might find between any two members of *Homo sapiens* today. There were probably many more species of hominin on Earth at any one time in the past 6 million years or so than *Homo sapiens* or its immediate ancestors. Second, and following from the arguments I've laid out in earlier chapters, the fossil evidence for such hominins will be meager. Third—and rather more controversially—there might be nonhuman species of hominin still around today, or which perished in historical times.

Let's look at these points in turn.

As we've seen, the preservation of creatures in the fossil record is vanishingly unlikely, particularly for those that lived on land. Hominins would have left a sparser fossil record than most, because they were always rare to start with. For all that, between around fifteen and twenty different species of hominin are known to have evolved since the hominin lineage split from that of chimpanzees. The number is inexact, partly because most of the fossils are very fragmentary, and also because scientists cannot always agree on the identification of any particular one, whether it belongs to a new species or is a member of a species that is already known.

Having read this far, it probably won't surprise you to learn that until recently the "picture" of human ancestry often relied on the assumption that members of one fossil species were directly ancestral to members of other species. This view still persists here and there, but there is one particular doctrine I wish to examine here—that is, the view that only one species of hominin could have lived on Earth at any one time.[7] The usual reason given for this idea was that the global ecology could only ever have had room for one species of hominin at a time. As soon as a new species of hominin appeared, the old one was inevitably driven to extinction. It follows from this view—which I'd call the this-town-ain't-big-enough-for-the-both-of-us hypothesis, except it's too unwieldy for everyday—that species of hominin had to be directly ancestral to one another, rather than cousins who shared a common ancestor in the past—there would have been no other source for a new species, other than the old one, already in existence. The scenario sprang, I think, from the conventional view of evolution as linear and progressive (the idea of ecological exclusivity being a scientifically dressed-up excuse for this, made after the fact) and from our own experience.

After all, every human being we know belongs to a single species, *Homo sapiens*. (There were once attempts to classify members of different human races as different species, but such work has long since been discredited and shown to be false.) As far as we know, no other hominin survives on the planet. From this it might be easy to assume that this situation was always so, yet the present era appears to be exceptional. As recently as 50,000 years ago, *Homo sapiens* shared the earth with at least five others—and these are only the ones we know about. Further back in time, when hominins were probably restricted to Africa, *Australopithecus* of various species coexisted with at least two species of early *Homo*. The idea that only one species of hominin lived on Earth at any one time was easy to accept when the fossil record of hominins was even worse than it is now. By 1976, however, the fossil record from east Africa, showing early *Homo* living alongside australopiths—could no longer be discounted. The human family tree was not a single line, but a bush with many branches, all but one leading to extinction.[8]

What this tells us is that our record of hominin fossils is important not by virtue of the fossils that have been found, because these are few, but by the oceans of ignorance that they punctuate. In almost all cases, newly found hominin fossils open up new vistas, new possibilities, that scientists had not imagined before the fossils were found. This tells us that the hominin record is not only sparse, but so sparse that even the general course of events in human evolution cannot clearly be discerned—much less a coherent narrative.

What else might lurk in the vast gaps between the tiny islets of knowledge represented by the few fossils that have been discovered?

The case of *Homo floresiensis* is particularly instructive. This discovery revealed the presence on a remote island of a peculiar hominin that had evolved in isolation for at least 100,000 years, and possibly more than a million, and whose anatomy spoke of an evolutionary divergence from the hominin line before the emergence of *Homo erectus*, or even the genus *Homo* itself.

The implications of *Homo floresiensis* for understanding the scale of our ignorance are immense. This single discovery showed that hominins might have migrated from Africa perhaps a million years earlier than anyone had thought, which means a million years of entirely unknown hominin evolution in Eurasia as yet completely undocumented by fossils, and of which everyone had been completely ignorant.

It showed that the usual scenario of human evolution, concerning the emergence of *Homo erectus* and its migration out of Africa around 1.9 million years ago, is based very much on our idea of human evolution as a narrative of progression, with scant regard paid to the poverty of the evidence required to support such a narrative.

Most of all, the discovery should prompt questions such as these: How likely do you think it is that researchers, excavating in just one cave on just one island in the vast archipelago that is Indonesia, just happened upon the one and only species of peculiar, endemic, primitive hominin that ever existed in Eurasia? And, given that *Homo floresiensis* lived until almost historical times, how likely do you think researchers just happened to have stumbled across the only species of archaic hominin to have survived to so late a date?

More likely, I think, is the alternative view, that the world was full of hominins of all kinds, some of them persisting until relatively recently, in geological terms. Given that fossilization is exceptional, especially for hominins, it would be no surprise if almost none of these species left any trace in the fossil record. The discovery of *Homo floresiensis* is proof enough, in the breach. If *Homo floresiensis* existed, then so must many others, in many other places.

This is not to say that the discovery of *Homo floresiensis* has not caused some scientists to take a new look at specimens that never quite seemed to fit into the conventional narrative. A puzzling skeleton from Nigeria, for example, was generally dismissed as an oddity: it looked archaic, but its owner lived in geologically recent times. Now Chris Stringer of the Natural History Museum in London and his colleagues think it might have represented a hitherto unknown kind of archaic human, surviving well into the era of *Homo sapiens*.[9] Meanwhile, a number of skulls of ancient hominins from China have defied categorization.[10] Early *Homo sapiens*? Not quite. *Homo erectus*? Not that, either.

The incredible growth of research into ancient DNA is beginning to shed some light on such matters. The sequencing of the genome of several Neanderthals shows that around 4 percent of the DNA in modern Europeans comes from that rugged acme of Ice Age cave life. More startling still was the sequencing of DNA from an otherwise unremarkable hominin finger bone preserved in Denisova Cave in southern Siberia.[11] The DNA signaled the arrival of a hitherto unknown hominin species, distinct from both Neanderthals and modern humans, which had lived in eastern Asia until as recently as 30,000 years ago. The latest known occurrence of a species as a fossil is never its last, so the Denisovans must have been around more recently than that. In a way they are still with us, because these archaic hominins have left traces of their genes in modern human populations throughout New Guinea and the western Pacific Ocean.[12] The discovery allowed a whole host of questions to be asked, questions whose framing had not hitherto been possible—were some of the enigmatic Chinese skulls from Denisovans? What about some of the strangely archaic-looking skulls of the earliest-known colonists of Australia?[13] Were they *Homo sapiens*, or perhaps Denisovans—or a mixture of both, or something else altogether? Largely thanks to Flores, the world of paleoanthropology (the study of fossil hominins) has, in the past decade, learned to appreciate the stark magnitudes of the unknown.

Now, to the third and perhaps most controversial of the three topics I raised above—is it possible that hominins other than *Homo sapiens* might still be living in the modern world, or, if extinct, perished only in historical times? The discoveries of *Homo floresiensis* and the Denisovans suggest that the question is not quite so outlandish as it might appear at first. After all, we know that the last known appearance of a species in the fossil record might antedate by some margin the actual date of a species' extinction.

The Denisovans are known—so far—from just one locality, so their time range is hard to estimate, but we can get some idea of the likelihood of *Homo floresiensis* persisting into the modern age. To recap: the skeleton of *Homo floresiensis* from Liang Bua cave, the best and most informative specimen of the species so far known, has been reliably dated to around 18,000 years ago. Other specimens of isolated bones, all from different layers in the same cave but attributable to the same species, range from 14,000 to perhaps as old as 95,000 years. Extensive evidence from elsewhere on Flores shows that hominins were making tools on the island for at least a million years. At the top end of the sequence, a layer of volcanic rock speaks of a massive volcanic eruption around 12,000 years ago. Layers deposited more recently show no sign at all of *Homo floresiensis* or of other creatures endemic to the island. Leaving aside local Floresian folk wisdom of the *ebu gogo*, the Little People who lived in the mountains (tales found pretty much everywhere), it is very likely that the volcanic eruption did for *Homo floresiensis* just as random, localized disasters have almost certainly tipped other, isolated hominins into extinction. It is a teasing thought, however, that had the eruption not happened, *Homo floresiensis* might have lasted into modern times, and, because of all the chances and mischances of life, death, and fossilization I've discussed in this book, it shouldn't be so surprising were other species of hominin to be found living obscurely on Earth with us.

Even today, when the earth teems with *Homo sapiens* wherever you look, and you'd think that scientists had shaken every tree and peered behind every bush on the planet, creatures as yet unknown to science emerge blinking into the light. Not just small, obscure creatures—insects, microbes, nematodes—but creatures large enough to do you an injury if they stepped on your foot.

Southeast Asia—home of *Homo floresiensis*—seems especially prone to this phenomenon. In 1937, a species of wild ox called the kouprey (*Bos sauveli*) was described for the first time, living in what is now Cambodia.[14] A species of the archaic fish known as the coelacanth was discovered living near the island of Sulawesi in 1998, a full 10,000 kilometers from the only other known species, in the Comoro Islands just off Madagascar,[15] itself a relic of an ancient lineage thought to have perished in the days of the dinosaurs until a recently dead one was brought ashore off South Africa in 1938.

In 1993, a report hit my desk in *Nature* of a large and very peculiar antelope, unknown to science, and described from horns and skins found in the huts of hunters working in the Annamite mountains in the remote border region between Laos and Vietnam.[16] The description contained no account of living animals, for none had yet been seen by scientists. It took several more years before the animal, by then known variously as the saola, the Vu Quang ox, or *Pseudoryx nghetinhensis*, was filmed, and very few have been captured and studied.[17] There can be few other cases of a creature that goes so far out of its way to go out of its way.

Such new discoveries do not always toil in obscurity, however—the okapi, a familiar zoo animal and friend of every player of Scrabble,[18] was discovered in what is now the Democratic Republic of the Congo as recently as 1901.[19] The discovery was not entirely a surprise, as rumors had been circulating among Africa hands for some time, particularly after the explorer Henry Morton Stanley (the same who presumed to have met Dr. Livingstone) described an as yet scientifically unknown ox-like beast in the 1880s.

The story of the okapi shows that it is possible for creatures of myth, rumor, and folklore to emerge into the light. Might the same be true of any of the several as-yet-mythological varieties of hominin believed by some to roam various corners of the earth? There is no reason in principle why it should not. However, virtually all the several species of large creatures described by scientists over the past century have been surprises—they were found without anyone setting out to look for them on purpose: in other words, rather in the manner of the coelacanth, the saola, and *Homo floresiensis*. It is perhaps significant, therefore, that the fabled Sasquatch or "Bigfoot" of North America, the Yeti of Tibet, the Orang Pendek of Malaysia, and other famous beasts (the Loch Ness monster being perhaps the most notorious) have failed to materialize despite decades of directed effort, often derailed by fairground hucksterism and not a few deliberate hoaxes.[20] The likelihood of an unknown animal being found appears to be inversely proportional to the efforts devoted to its pursuit. This doesn't mean that *Homo sapiens* is necessarily the only hominin to survive—only that news of any others will come suddenly and quite unexpectedly. *Homo floresiensis* was unexpected—and much the same can be said of almost every other discovery made concerning human evolution. Such findings usually challenge deeply held beliefs about human uniqueness that are very hard to shift. It is perhaps not surprising, therefore, that almost the only discovery of a fossil hominin that immediately convinced all the experts turned out to have been a fraud.

The first fossil hominin to be recognized as such was dug up in Gibraltar in 1848, but the finding that really marks the start of paleoanthropology was made in 1856, three years before the publication of *The Origin of Species*. This was a skullcap and bones collected in a cave high in the Neander valley near Düsseldorf in Germany.[21]

The findings were immediately a source of controversy. Those who thought that the creature was an extinct kind of human were brave indeed, given that "extinction" as a concept was still quite new. Nowadays the findings are regarded as having belonged to Neanderthal Man (*Homo neanderthalensis*), perhaps the closest extinct relative of *Homo sapiens*. The Gibraltar skull is also regarded as Neanderthal. At the time, though, the authorities of the day thought that the Neanderthal finds belonged to a modern human, if of perhaps a primitive sort, and possibly suffering from rickets, a vitamin deficiency that results in deformities of bone growth. It was also suggested that the finds belonged to a Mongolian Cossack who'd been involved in the Napoleonic Wars. These views brushed aside the obvious problem that the poor soul who gave up his bones to posterity, injured and perhaps deformed, scaled the sheer walls of a cliff to find a convenient cave in which to expire. In that way, paleoanthropology started as it was to go on—by loud proclamations from establishment alpha males that the latest discovery is in fact that of a diseased or deformed member of modern humanity. Either that, or it's an ape. Critics rarely allow that the new discovery might be anything that might genuinely expand our vision of what is known.

Darwin had the percipience (in his book *The Descent of Man*) to suggest that as the closest still-living (as opposed to extinct) relatives of modern humans were chimpanzees and gorillas, which lived in Africa, then the deepest roots of humanity lay in that continent. This seems entirely logical to us, in hindsight, given almost a century of African fossil discovery, but it was not always so. At the end of the nineteenth century and well into the twentieth, it was commonly thought that Asia, not Africa, was the cradle of humanity. One man, a Dutch doctor named Eugène Dubois, was so convinced of this that he staked his career on it, traveling to Java in the (then) Dutch East Indies to search for fossil evidence that might bear on human evolution. By the most amazing luck, he found it. In 1891 a skullcap, and later, limb bones, came to light, which Dubois named *Pithecanthropus* (ape man). Further finds of this creature were made in subsequent years.[22]

Pithecanthropus showed an interesting combination of somewhat humanlike limb bones but a skull rather smaller than that of modern humans. *Pithecanthropus* could be seen as an intermediate between apes and humans, but ran against the prevailing theoretical view that as modern humans are distinguished by big brains, then it must have been the case that big brains evolved first, before humans learned to stand fully erect. As the musician George Clinton once memorably put it in another context—free your mind and your ass will follow. *Pithecanthropus* was therefore generally seen as an apish side issue, perhaps a giant gibbon, but not anything especially close to human ancestry. Thus was a mark made for the second of the two conventional reactions to new discoveries of members of the human family: if it's not a diseased human, then it must be some kind of ape.

The most important and influential fossil hominin discovery ever made was a skull and jaw of a fossil human from a gravel pit in Piltdown, in southern England, in 1912.[23] The skull was undeniably old, as shown by fossils of ancient mammals found in the same gravels, but it looked remarkably modern. The jawbone, assumed to have been associated with the skull, looked very apelike, with a receding chin. The dynamite combination of modern-looking skull and primitive jaw showed that the brain had, indeed, led the way in human evolution, dragging the brutish body after it.

It might be no coincidence that the heyday of Piltdown Man coincided with the nadir of Darwinism. Piltdown came at just the right time to fall victim to the vacuous storytelling as condemned by scientists such as William Bateson. If distinguished anthropologists, who ought to know, assure us that natural selection drove the evolution of the brain before all else, then, in short, it did, and that was that. *Eoanthropus* was orthodoxy for a generation, the fossil against which all others would be judged.

Eoanthropus had had a clear run for more than a decade when the first evidence for human evolution in Africa turned up. The discovery threatened to overturn the now-established view of the "Piltdown committee"—the London-based group of grandees whose opinions on human evolution went unchallenged.[24] The report came from Raymond Dart, an Australian-born medical scientist at the fledgling University of the Witwatersrand in South Africa. Dart had been sent a crate full of fossils blasted out of a lime works at Taung in the Transvaal. One of the fossils was the cast of a brain of a small, apelike creature. Another was the skeleton of the face that hafted onto the brain cast, as snugly as a cricket ball in a wicket keeper's mitt.[25] Dart, who had been taught neuroanatomy in London by another expat Australian, Grafton Elliot Smith—one of the Piltdown committee—immediately grasped the significance of the find. This was not an ape, but a child of a new species, intermediate between apes and modern humans. Dart sent a preliminary description of the find to *Nature*, where *Australopithecus africanus* (southern ape from Africa) was published.[26]

The reaction to the Taung "baby" was immediate and negative. Letters sent to *Nature* from the various members of the Piltdown committee, including Dart's mentor, Grafton Elliot Smith, damned Dart's finding with faint praise.[27] Yes, the finding was important, but fossils of juveniles are always hard to judge, they said. Humans and chimps look far more alike as babies than they do as adults, so the Taung baby could be an infant ape just as well as an infant ape-man. The Piltdown committee effectively suppressed the publication of Dart's subsequent monograph, which contained much evidence in favor of the ape-man hypothesis that was either incompletely considered or not available when Dart sent his short communiqué to *Nature*. In 1929, the Royal Society in London rejected the monograph for publication—the referees certainly included members of the Piltdown committee—and the manuscript lay ever afterward buried with Dart's papers at the University of the Witwatersrand. Would paleoanthropology have been changed had the monograph been published? It's impossible to say.

Dart's salvation came in the form of more fossils to back up his ideas. Robert Broom—a paleontologist and a rather more intrepid character than Dart—saw the Taung fossil for himself, and reported to *Nature* that it was precisely as Dart had said it was.[28] Although Broom had had the advantage over the Piltdown committee of actually having seen the fossil for himself, it took Broom's discoveries of several more fossil hominins[29] from cave deposits in South Africa for the idea of ape-men to take hold.

Broom's hominins were a varied lot. Although no more fossils were forthcoming from Taung, fossils of what looked like adult versions of *Australopithecus africanus* came from an ancient cave called Sterkfontein, whereas fossils of a rather different creature, eventually called *Australopithecus* (or *Paranthropus*) *robustus*, emerged from other sites, Swartkrans and Kromdraai. *A. africanus* was slightly built, with somewhat humanlike teeth not specialized for any diet in particular. *A. robustus*, in contrast, had massive jaws and big, blocky teeth, perhaps more suitable for a diet of tough vegetation such as roots, seeds, and nuts. Broom's almost single-handed barrage of papers on these creatures to *Nature* from the mid-1930s onward was largely responsible for rehabilitating Dart's reputation.[30]

The variety of these finds should have been evidence enough that the human family was diverse, and that more than one kind of hominin existed at any one time: the first evidence that human evolution was uncertain and bushy, far from the single lineage I caricatured in chapter 1. At the time, however, the geological ages of these fossils was not known with any certainty. All the australopiths came from cave deposits, which are invariably a jumble of things that have fallen in, or were brought in by predators at various times. Even today, getting reliable dates for fossils found in caves is a difficult business, and that's with the battery of modern techniques for dating that hadn't been invented in Broom's day.

As more and more fossils came to light, it became clearer that the australopiths did not conform to the George Clinton model of evolution. They had rather small brains, but would have walked erect—the precise opposite of the model espoused by the Piltdown committee.

Further evidence came from China in the late 1920s and early 1930s, at a site called Dragon-Bone Cave (Chou Kou Tien, modern-day Zhoukoudian), where a Canadian called Davidson Black and his colleagues reported fossils of a creature they called *Sinanthropus*.[31] "Peking Man" had a larger brain than *Australopithecus*, but smaller than modern humans.[32] *Sinanthropus* was associated with stone tools, and perhaps even the controlled use of fire—hallmarks, it was thought, of technology, and therefore of humanlike activity.[33] *Sinanthropus* was later shown to be very similar to Dubois's *Pithecanthropus* from Java,[34] and the two were united into one species, *Homo erectus*—the Man who stands upright.

By the late 1930s, a picture of early human evolution was beginning to emerge that has remained intact, more or less, ever since. The earliest members of the human family evolved in Africa, and were typified by forms such as *Australopithecus*—rather apelike, with small brains, but which nevertheless walked upright. Later on, hominins dispersed into Eurasia, acquired tools and a certain stature, and became *Homo erectus*—with a brain larger than those of australopiths, but smaller than in modern humans. *Homo erectus* walked upright. His name said as much.

The steady accumulation of evidence made Piltdown Man look like an increasingly anomalous side issue, ever harder to fit into evidence that challenged the preconceptions of the experts. Big-brained Piltdown might have had some support from Neanderthal Man, which had, if anything, a larger brain than seen in modern humans—an inconvenient fact that is usually brushed aside in the canonical picture of acquisition and improvement. Neanderthals were also seen as stooped and shambling, as Piltdown was meant to have been. But the skulls of Neanderthals, while large, are very distinctive, and quite different from those of modern humans. And the picture of Neanderthals as stooped comes from the interpretation of just one skeleton, of an elderly male crippled with arthritis. In reality, Neanderthals stood as erect as any healthy modern human. Piltdown stood alone.

Eventually, the penny dropped. As the years wore on, it became ever clearer that Piltdown was not so much anomalous as embarrassing. In 1953, proof came of what many had already come to suspect, that Piltdown Man was a fraud.[35] The skull looked like that of a modern human because it was one. The apelike jaw had come from an orangutan. The joint where the jawbone would have attached to the skull had been broken, so nobody could have seen that the two didn't fit together. The teeth in the orangutan jaw had been filed down so that they didn't look so apish to have given the game away, and the whole arrangement had been stained to make the bones look very old. The gravel pit at Piltdown had been salted with bones of archaic mammals from elsewhere.

The identity of the hoaxer remains unknown to this day. There has been some suggestion that it was one Martin Hinton,[36] an expert on fossil rodents at the Natural History Museum, who had the means and the technical knowledge, and also a motive: a grudge against Arthur Smith-Woodward, his boss, a prominent paleontologist—and a leading light on the Piltdown committee, and a critic of Dart's *Australopithecus*.

Whoever was responsible, the joke went far too well—perhaps so well that there was no possibility of a safe confession for the hoaxer. Smith-Woodward and his cronies bought the story without question. Further "finds" at Piltdown relating to the "First Englishman" included a hunk of bone deliberately carved into the shape of a cricket bat, clearly meant to be so ridiculous that someone, surely, would have suspected something. This, too, was treated as genuine.

At the risk of laying it on with a trowel, the moral is that it is very easy to see fossil evidence (or, indeed, any scientific evidence) through the highly selective and distorting lenses of one's deeply held preconceptions, rather than for what it plainly is. If I have gone on about it at some length, that's because it can be seen as the message for this whole chapter, that the discoveries made in the course of shedding light on human prehistory have a habit of challenging preconceptions—and indeed for this whole book, that when looked at dispassionately, many if not all the attributes we think of as distinctly human are in fact nothing of the kind, and even if they are unique to humans, this uniqueness is in itself nothing special.

With the final unraveling of Piltdown the focus moved back to Africa, and the name of Leakey. Louis Leakey, the son of a missionary who came to preach the gospel to the Kikuyu, soon became fascinated with the search for what he called in a later book "Adam's ancestors."[37] The search was long, hard, and, for thirty years, mostly fruitless. In 1959, however, Leakey's wife, Mary, discovered the skull of a fossil hominin at Olduvai Gorge in what is now Tanzania.[38] This creature was *Zinjanthropus boisei*, a robust australopith, similar to *Australopithecus robustus* from South Africa. (These days many paleoanthropologists prefer to group all robust australopiths together in a separate genus, *Paranthropus*, so that *A. robustus* is *Paranthropus robustus* and Leakey's "Zinj" is *Paranthropus boisei*.)

In 1964, Leakey announced the discovery of what he claimed to be the earliest evidence for the genus *Homo*. This was *Homo habilis* (handy man).[39] The name was attached to a second hominin discovered at Olduvai, less robust than "Zinj."

Although Leakey's 1964 paper describing *Homo habilis* was very careful, laying out technical statements on the anatomy of the genus *Homo* in general and the species *Homo habilis* in particular, the presumption was clear: if *Homo habilis* and *Zinjanthropus* were at the same site, associated with stone tools, then *habilis*, with its larger brain, was probably the toolmaker. "When the skull of *Australopithecus* (*Zinjanthropus*) *boisei* was found on a living floor at F. L. K. I," wrote Leakey and colleagues,

> no remains of any other type of hominid were known from the early part of the Olduvai sequence. It seemed reasonable, therefore, to assume that the skull represented the makers of the Oldowan [stone tool] culture. The subsequent discovery of remains of *Homo habilis* in association with the Oldowan culture at three other sites has considerably altered the position. While it is possible that *Zinjanthropus* and *Homo habilis* both made stone tools, it is probable that the latter was the more advanced tool maker and that the *Zinjanthropus* skull represents an intruder (or a victim) on a *Homo habilis* living site.

Because you cannot reliably infer the behavior of an extinct creature from its bones, the whole definition of that creature, if encountered as a fossil but classified according to its presumed behavior, becomes debatable. The fossils we have are fragmentary. One debate centers on whether there is one species of early *Homo*—*Homo habilis*—or two, the other being *Homo rudolfensis*, a name attached to a skull discovered in

1972 by Bernard Ngeneo—a member of the research team led by Leakey's son, Richard—at Koobi Fora on the eastern shore of Lake Turkana in Kenya.[40] A third species of early *Homo* was recently added to the mix. This is a skull discovered in 1977 from Sterkfontein, a site famous for its australopiths. The skull was originally assigned to *Homo habilis* but has now been renamed *Homo gautengensis*.[41]

The problem is that the type specimen of *Homo habilis*—that is, the fossil used to name the species—is a jawbone with teeth, and the type specimen of *Homo rudolfensis* is a skull, complete in most respects except that it lacks a jaw and teeth. This means that there is no way to compare the two species directly, so any decisions about the status of these species has to be made in a roundabout way, by comparing them with other fossils that might very well have their own problems of interpretation.[42] The confusion deepens with the possibility that any and all early *Homo* species might really be australopiths, and not *Homo* at all. The murk is thickened by a general fogginess about what features make a hominin *Homo*—but mostly by the general lack of fossil evidence. Even after almost a century of sustained effort, the fossil record of hominins is too slender for us to say anything definite about the origins and general characteristics of the earliest members of our own genus, *Homo*.

Some scientists, notably Bernard Wood of George Washington University in Washington, DC, have suggested that these early species of *Homo* look sufficiently archaic—and too much like australopiths—that to include them in the genus *Homo* makes defining our own genus even more difficult than it is already.[43] Wood prefers to cast these hominins as australopiths, similar in many ways to *Australopithecus africanus*—a species that has also been hard to define, given that its name was originally coined to refer to a baby, rather than an adult in which the full expression of a species' distinctive traits might be seen. The situation has been complicated further with the detailed description of two partial skeletons of an australopith from Malapa, a cave near Sterkfontein.[44] This hominin, named *Australopithecus sediba*, had a small brain, but features of its skeleton are reminiscent of early *Homo*.

What of the tools? Evidence for tools now goes back at least 2.5 million years—for tool use, perhaps as long as 3.39 million years, if the scratches seen on animal bones excavated from a site in Ethiopia were deliberately made by hominins.[45] This makes toolmaking far more ancient than the genus *Homo*, even if *Homo habilis* is admitted to the club.

It is possible, indeed likely, that australopiths made tools. If *Homo floresiensis* is the descendant of an australopith, rather than *Homo erectus,* then the case is made, given that tools have been found associated with Flo.

The error is to construct an argument that is both circular and spurious. If we assume that only members of the genus *Homo* can make tools, then anything associated with stone tools must be in the genus *Homo.* This ignores the possibility that the tools might have been made by Zinj or indeed any other hominin, including species as yet undiscovered. Now that we have good reason to think that australopiths indeed made tools, the necessary restriction to the genus *Homo* becomes nonsense. The alternative is to admit australopiths to *Homo,* which would then make *Homo* even harder to define than it is already.

Worse, though, is the conceit that toolmaking necessarily accompanies a bigger brain, such that when a brain becomes big enough, facilities such as technology become possible. As I show later in this book, many animals with brains much smaller even than that of *Homo floresiensis* make and use tools. Conversely, organisms as simple as bacteria can make structures at least as elaborate as stone tools, but nobody would accuse such creatures of having any brains at all. Leakey, like the Piltdown committee before him, was in danger of making unwarranted assumptions about the progressive evolution of hominins where no such assumptions were justified.

After Leakey, paleoanthropology split into two streams, one going further forward in time, the other, backward.

Forward first. With *Homo erectus* in Asia and Neanderthals in Europe, but *Australopithecus* and very early *Homo* in Africa, the consensus view emerged that *Homo* migrated out of Africa with, or soon after, the evolution of *Homo erectus,* around 1.8 million years ago. Further hominins found in Eurasia appeared to confirm this view.

Recent discoveries include several spectacular specimens of hominin skulls and skeletons in caves in the Sierra de Atapuerca in northern Spain.[46] These finds, while remarkable, are not alone. Remains of hominins have been found across Eurasia from Britain to China. Some appear to belong to *Homo erectus,* whereas others are much harder to place, and are conventionally lumped into a kind of dustbin called *Homo heidelbergensis,* named after a mandible discovered in Germany in 1907, and conventionally regarded as a generalized Eurasian form whence descended the Neanderthals, and possibly also *Homo sapiens,* if enigmatic finds from Africa such as *Homo rhodesiensis* (a distinctive skull found in 1921 in what is now Zambia) belongs to this increasingly inclusive transitional form. Another species, *Homo antecessor*, comes from deposits at Atapuerca thought to be older than those yielding the bones of supposed *H. heidelbergensis*.[47]

This conventional view has run into problems. The first is the nature of *Homo erectus* itself. Did this species really originate in Africa? Finds assignable to this species have indeed turned up in Africa, notably a near-complete skeleton of a youth discovered at Nariokotome on the shores of Lake Turkana in Kenya in 1984, by the legendary fossil hunter Kamoya Kimeu, one of Richard Leakey's "hominid gang."[48] The youth clearly belonged to *Homo*, based not just on the features of the skull, but on the skeleton, which had the cylindrical ribcage and long legs seen in *Homo*, rather than the more conical ribcage and shorter legs typical of *Australopithecus*. Some researchers, though, found sufficient differences between the Nariokotome skeleton and some other African finds and later *Homo erectus* to create a new species, *Homo ergaster*, to encompass early *erectus*-like hominins from Africa.[49] Very early examples of hand axes, a style of stone tool very much associated with *Homo erectus*, have also been found in Ethiopia and Kenya[50]—with, of course, the usual health warnings about linking tools and their makers.

So, what's the problem? Here's the deal: *Homo erectus* evolves in Africa around 1.8 million years ago; evolves advanced hand axes rather than the simpler pebble tools of *Homo habilis* (and maybe *Australopithecus*); colonizes Rest of World. To me, this narrative seems rather too biblical to be credible.

The first crack in the story was the discovery of a remarkable collection of hominin skulls and bones from rocks beneath a medieval monastery at Dmanisi in the Republic of Georgia, associated with primitive tools, and dated to between 1.85 and 1.78 million years ago.[51] The Dmanisi hominins are arguably the oldest known hominin fossils outside Africa—but if they are the descendants of the first bold exiles from Africa, they seem to have taken a step backward. They are sufficiently primitive to have drawn comparisons with *Homo ergaster* rather than *Homo erectus* but have also, lately, acquired their own species name, *Homo georgicus*.[52]

It's very tempting to view the Dmanisi hominins as primitive members of *Homo erectus* (or something closely related to it) caught in the act of migrating out of Africa. Such temptations should be resisted. To be sure, there are bits and pieces of *Homo erectus*, and of stone tools indicative of their passing, recovered from all over the Old World that could quite plausibly be strung together to make a story of migration from Africa, but that would be to ignore the gaps in time and space that must be bridged. Rather than *Homo erectus* having evolved from an earlier form of *Homo* in Africa and moved into Eurasia, it is perfectly possible for *Homo erectus* to have evolved in Asia from some even earlier form and migrated back into Africa, replacing *Homo habilis*. The Dmanisi hominins might therefore be creatures caught in the act of coming home, not venturing forth. This scenario might seem a little contrived, especially as no fossils of hominins older than 1.7 million years are currently known from outside Africa. But the dates for many early hominins in Europe are constantly being pushed backward toward the 2-million-year mark. And the existence of *Homo floresiensis*, which looks arguably more like *Australopithecus* than a dwarfed *Homo erectus*, suggests that hominins left Africa much, much earlier than had been thought possible.

The recent discovery of early hand axes together with pebble tools from the western shore of Lake Turkana in Kenya further blurs the picture.[53] It suggests that the first hominins to have left Africa might have fled without their distinctive hand axes. It also raises the possibility that the first exiles were more primitive than *Homo erectus*—and the even more remarkable possibility that *Homo erectus* didn't evolve in Africa at all, but having evolved in Asia—perhaps from even earlier African roots—went back to Africa again.[54] Everyone knows that the Israelites crossed the Red Sea, but nobody said anything about them coming back.

Rolling the tape forward, from about 1.8 million to 200,000 years ago, we see in the fossil record the first signs of behavior that seems distinctively human, as opposed to just hominin. The earliest known remains attributable to *Homo sapiens* are almost 200,000 years old and come from Ethiopia.[55] At about the same time, cave sites in South Africa show signs of new things, such as shells pierced to make ornaments, the use of a natural pigment called ocher in decoration, and the extensive exploitation of seafood.[56] When *Homo sapiens* evolved, the first thing it did was head for the beach.

The story goes that once modern humans evolved in Africa, they spread throughout the world, displacing any other hominins they might have come across, most notably Neanderthals in Europe. This "out-of-Africa" tale was vigorously countered by another view, called "multiregional continuity," that *Homo sapiens* evolved several times, independently, from various forms scattered throughout the world, including Neanderthals. At this point I shall point you to figures 5 and 6 in chapter 1 and ask you whether you think, on the basis of the fossil evidence, either view stands up.

The out-of-Africa idea received a huge boost in 1987 with a study on human evolution that broke new ground by not being based on sparse, fragmentary fossil evidence, but on comparisons between people alive today. Our inheritance is encoded in the genetic material, DNA, almost all of which is found in the nucleus of each cell. But cells also contain other bodies, called mitochondria, which have DNA. It so happens that this mitochondrial DNA (or mtDNA) is passed strictly down the female line. Writing in *Nature*, the late Allan C. Wilson and his colleagues described how they analyzed the mtDNA from 147 people of diverse origins and used the pattern of similarities and differences between the samples to sketch a kind of evolutionary genealogy of humanity.[57] The results showed that the greatest diversity of mtDNA was to be found in Africa, and that mtDNA from everywhere else seemed to have been an offshoot of an ancient African form. This idea gave credence to the view that modern humans evolved in Africa and spread throughout the rest of the world. Calculations of the rate at which mtDNA would acquire new variations suggested that humans left Africa roughly 200,000 years ago. Given that mtDNA is passed down exclusively from mothers to daughters, the authors wrote that "[a]ll these mitochondrial DNAs stem from one woman who is postulated to have lived about 200,000 years ago, probably in Africa." The significance was not lost on the author of an accompanying commentary in *Nature* entitled "Out of the Garden of Eden," which described the Wilson paper as reporting that "Eve was alive, well, and living in Africa around 200,000 years ago."[58] With Genesis in your PR department, you can hardly go wrong. It's perhaps fortunate that subsequent work has largely borne out the idea that modern humans originated in Africa.

The picture has, perhaps inevitably, become more complicated. More recent studies on nuclear DNA, including DNA recovered—remarkably—from Neanderthals, shows that modern humans didn't completely replace earlier species of hominin. If you are of European descent, then around 4 percent of your genes came from Neanderthals.[59] If you're from New Guinea, you might boast an even more remarkable heritage, for some of your genes come from the still-obscure Denisovans.[60] In his book *The Origin of Our Species*, Chris Stringer suggests that some archaic hominins might have survived even in Africa until a few tens of thousands of years ago. This idea is supported by genetic work showing traces of interbreeding with ancient hominins as yet unknown in some modern African populations.[61] Everywhere you look, we all bear some genetic traces of hominins past.

In any case, it must be remembered that "mitochondrial Eve" was not the only woman around at the time, only the one whose mitochondrial DNA appears to have survived until the present day. It is, perhaps, no more than luck that it was she, rather than any other female, living earlier or later, who turned out to have been the ancestor of all the mtDNA found (so far) in modern humans.[62]

The tale of out-of-Africa makes a piquant contrast with the story of Piltdown. With Piltdown, a fossil that accommodated prevailing prejudices about the course of human evolution was found to be a fake. When more (indeed, when any) evidence surfaced, it was shown that hominins walked upright before they got bigger brains. This remained true despite the squabbles about which of the fossils should be called *Homo* and which *Australopithecus*, and indeed over which one was related to whom.

Out-of-Africa based its imagery on biblical narrative, and turned out to have been correct. This needn't have been the case. The Wilson study was based on rather few samples, and, as it happened, the African sample came from African Americans, some of whose ancestries might not have been purely African. One of the functions of science is to test new results and, if possible, extend them with new data. It so happened that further work tended to support rather than refute Wilson's conclusions. Importantly, we should not take that outcome as a given. In which case, the invocation of Eve and the Garden of Eden was brave indeed, and could be interpreted as grandstanding ahead of the evidence.

The Piltdown committee looked at the evidence and asserted that it supported their particular view of how human evolution ought to have been. That they had been fooled by a forgery shows only how strong such received notions can be, and how hard they are to shake. The Wilson paper came out in support of the out-of-Africa view. Although it received pretty much immediate, universal approbation, the same might be said of Piltdown. Science is not a democracy: public acclaim is itself no guarantee that any view is the correct one. The co-option of the biblical narrative was in its way as prejudicial as the Piltdown committee's view that brains came before bipedality. In the end it all comes down to chance. The Piltdown committee was (completely) wrong; the out-of-Africa view was (mostly) right. But as I showed in chapter 1, the "true" story of human evolution has no obligation to cleave to any story we might imagine, no matter how informed our guesswork.

And now, backward.

For a long while, the most ancient known hominin was *Australopithecus africanus*, reckoned to have lived from about 3 million years ago. However, it was clear that the hominin line was much more ancient than told by the fossils. First, australopiths share more traits with humans than they do with apes: there was still room for hominins more primitive than australopiths, between *A. africanus* and the latest common ancestor of chimps and humans. Second, estimates of evolutionary rate based on fossils suggested that the human lineage split from that of chimps 10–15 million years ago. This estimate was shortened dramatically when it became possible to estimate evolutionary rate directly from DNA differences between modern humans and chimps. The consensus now is that humans and chimps diverged between 5 and 7 million years ago.

Before that was an uncomfortable gap. Between around 5 and 10 million years ago, the fossil record was until recently almost completely blank. This was particularly frustrating, as it is in this interval that the lineage leading to hominins is thought to have diverged from that leading to chimps. Yet the fossil evidence for this most epochal, most defining event in human evolution, in which many generations of the earliest chimps and hominins lived, died, and eventually went their separate evolutionary ways, might be fitted into one rather small box.

In the 1970s the focus of fossil exploration moved northward from Kenya to Ethiopia, where rocks of the right age came to light.[63] Deposits exposed by the evolving Awash and Omo rivers, and in the Afar depression, provide a rich history of the Rift Valley older than Olduvai Gorge, the Turkana Basin, and the caves of South Africa. After much work, hominins began to turn up, the most famous being a partial skeleton of a diminutive female hominin known to her discoverers as "Lucy" and officially described as *Australopithecus afarensis* (southern ape from the Afar region).[64] Lucy was around 3.6 million years old. She was clearly more primitive than other hominins so far known, but she was definitely a biped. There was as yet space for even older hominins to be found.

An even older and more primitive creature, *Australopithecus* (now *Ardipithecus*) *ramidus* was discovered in the early 1990s, again in Ethiopia.[65] The remains were fragmentary—mostly scraps of teeth and jaws—but there was just enough to suggest that the fossil record of hominins stretched back 4.4 million years. The fragments were accompanied by a skeleton, so fragile and crushed that the fossils took more than a decade to be prepared from the concrete-like rock matrix.[66] Ardi was like Lucy, only more so: very likely a biped, but small, primitive, and somewhat apelike.

Ardipithecus ramidus is now joined by a number of other very early east African hominins known from fragmentary remains—*Australopithecus anamensis*, *Ardipithecus kadabba*, and *Orrorin tugenensis* extend the hominin record back beyond the 5-million-year mark.[67]

The further one travels back in time, the more fragmentary the fossils become—and the harder it is to distinguish them as hominin. This is only to be expected when one traces an evolutionary lineage backward. First one distinctive human trait disappears, then another, until one is left with a blank canvas of an ancestor on which one might paint any picture one likes. This exercise reminds me of one of my favorite playground jokes, which, like all such things, is more profound than it seems at first.

> Q: Why is an elephant large, gray, and wrinkled?
>
> A: Because if it were small, round, and white it would be a Ping-Pong ball.

To complicate matters, the very earliest hominins did not exist for our retrospective convenience. They, like every other creature, played their part on the tangled bank of Darwin's imagination, living alongside and competing with many other species. If the latest common ancestor of chimps and hominins had no features pointing to either one or the other in particular, it would certainly have had traits all its own that were lost in both subsequent lineages.[68] Even if you held in your hands the skull of the latest common ancestor of apes and humans, you could never know that you had done so.

In 2001 I had just such an experience. When I heard that Michel Brunet and his colleagues from the University of Poitiers in France had discovered something interesting, I invited myself to Poitiers to see the evidence in person. A few days later, Brunet put a fossil in my hands and left me alone in a room with it to contemplate the long evolution of humanity.

The fossil was a complete skull excavated from what was once a lush lakeshore, but now a blasted desert in the central African country of Chad. About the size and weight of a house brick, the skull was somewhat crushed, but it was far more complete than most hominin fossils of any age. The age of this skull was uncertain, but based on the many fossils of archaic mammals accompanying the find, it was believed to be somewhere between 6 and 7 million years old—right around the time when the human and chimpanzee lineages are thought to have diverged. Brunet and colleagues later described the skull in *Nature* as *Sahelanthropus tchadensis* (Sahel Man from Chad),[69] asserting its hominin status based on details of the dental anatomy and the position of the foramen magnum—the hole in the base of the skull that admits the backbone and spinal column. A foramen magnum at the back of the skull suggests that the owner was a quadruped, like a chimp. A foramen magnum set well beneath the skull, in contrast, suggests that the owner walked upright, like a hominin. The situation in *Sahelanthropus* seemed to suggest at least a tendency toward bipedality.[70]

Criticism of *Sahelanthropus* was swift,[71] and very reminiscent of the early comments aimed at Dart after his publication of *Australopithecus africanus*. *Sahelanthropus* couldn't be a hominin, but an ape, more specifically a female gorilla. Further work has gone on to show that *Sahelanthropus*, while apelike in many ways, is quite distinct. Of course, trying to work out whether a creature so close to the common ancestry of chimps and humans is more closely related to humans or to chimps will be practically impossible, as any distinctive features it has might be purely idiosyncratic, and have nothing to say about later evolutionary history. It could even be that *Sahelanthropus* branched off the evolutionary tree before the ancestors of humans and chimps became distinct. If *Sahelanthropus* was a biped, it could be that the common ancestor of humans and chimps was a biped, and that chimps later lost this facility.

If *Sahelanthropus* was destined to become an elephant, it made a very fine Ping-Pong ball.

To hold a fossil such as *Sahelanthropus* in one's hands is not to experience Schliemann's joy, when excavating a Mycenaean death mask, of coming face to face with Agamemnon. One does, however, find oneself, looking into those blank and squinting eye sockets, and asking questions of the boundaries of knowledge. What are we to make of this skull, almost the only evidence for the existence of a hominin (if that's what it is) to have been unearthed from the otherwise yawning void that stretches between around 10 million years ago (when the world teemed with fossil apes) and around 5 million years ago, when *Orrorin* and *Ardipithecus* start to appear? As the nineteenth-century Scottish preacher and geologist Hugh Miller wrote in his book *The Old Red Sandstone* (1841), the questions we would most like to ask fossils might forever remain unanswered: "We cannot catechise our stony ichthyolites, as did the necromantic lady of the Arabian Nights the coloured fishes of the lake, which had once been a city, when she touched their dead bodies with her wand, and they straightway raised their heads and replied to her queries. We would have many a question to ask them if we could—questions never to be solved." Instead, one ends up talking only to oneself, mostly about how little we know about anything, and of how vast is the ocean of ignorance in which we flounder.

It says something about the hominin fossil record that the discoveries of *Orrorin, Sahelanthropus*, and *Ardipithecus* have all been made very recently, in the past twenty years, and thanks to titanic efforts from scientists from many different countries, not least from the countries in Africa where the fossils are found.

Given the extent of our ignorance, and how it increases with every passing day, one can only be surprised—retrospectively, of course—about how loudly and certainly paleoanthropologists of the past asserted that the course of human evolution went this way or that, when the evidence was so sparse that practically any course of evolution might have been possible. In which case it is no surprise that they were often completely wrong, and if they were right, it was as much by luck as by judgment.

6: *The Human Error*

We human beings have at least two remarkable abilities.

One of them is the ability to recognize patterns. This facility is, in fact, the basis of all science: pattern recognition is the foundation of all knowledge and understanding, for without some ability to compare and contrast the properties of objects, no order can ever be discerned. Whether mystical nature philosophers or hard-headed Darwinians, all biologists recognize that within the apparent riot of biological diversity one can see a clear pattern, that of a tree, and from that, one can begin to ask how such a pattern might be generated.

Pattern recognition is likewise the basis of paleontology. The best fossil hunter in the Gee family is Gee Minima, who from an early age has been able to pick out fossil sea urchins from the jumble of rocks and detritus on Cromer beach. More than any other family member, she can see the distinctive double rows of dots that signal the positions of the tube feet of an animal that lived here more than 70 million years ago. Now she applies her keen eye for design and detail to the arts, fashion, and textiles, and can pick out details of shape and line that her sartorially challenged father plainly misses.

It is easy to see why pattern recognition is such an asset. Without some ability to categorize objects, the complex and crowded world in which we live would indeed be a dangerous place. In earlier times, an ability to recognize and quickly infer the nature of an approaching object, without taking time to explore it first, might have been a lifesaver. Those early hominins unable to tell the difference between a dead branch and a black mamba, or who misinterpreted the growl of an approaching leopard as the purr of a cuddly kitten, would stand less chance of passing on their genes to the next generation than those who saw the patterns, sorted these objects into the right categories, made the right choices.

And so we use that ancient circuitry today, and every day, when trying to make sense of our world. In his book *Us and Them* (which is probably the best anthropology book I have ever read, and if you haven't read it, I implore you to do so), David Berreby notes how this ability prompts us to leap to snap judgments about our fellow humans that on closer inspection they might not deserve. That rowdy crowd of tattooed and pierced bikers hanging around in your favorite restaurant? Instinct and experience—if only perhaps vicarious—might make you turn on your heel and walk out for fear of being mugged. The stories you've heard . . . Your instincts might well be right, and could save your life. Except that further investigation might have revealed that these particular bikers are all college graduates devoted to their mothers, have congregated to celebrate their charity bike ride to raise money for a sanctuary for abandoned puppies, and have chosen this restaurant because they've heard that the chef cooks up a crème brûlée that's to die for.

First impressions can save your life—or tell you lies.

So, while we are very good at recognizing objects, our talent is so refined that we are inclined to see patterns where there aren't any. Almost everyone who looks at the surface of the moon sees a human face, even though we know quite well that the features responsible for the illusion are in fact gigantic plains of ancient lava, and nothing to do with faces at all. We are perfectly aware of our tendency to make nonexistent connections, to spot nonexistent patterns. In such error lies much of importance and interest in our cultural heritage, in images of all kinds from classical trompe l'oeil to surrealism, in the comedies of errors in Shakespeare's plays and Mozart's operas to the cheapest farces, and in just about every joke you can think of. Here is an example (I have better ones, but they are too rude for a family audience).

> A: I say, I say, I say, how do you tell the difference between a postbox and the back end of a cow?
>
> B: I don't know, how *do* you tell the difference between a postbox and the back end of a cow?
>
> A: Well, if you don't know the answer to that, I won't send *you* to post the letters.

"Ba-boom," and, moreover, "tish."

So, much as we might indulge children who see elephants and railway trains in passing clouds, not to mention scoff at people who see images of Jesus in pieces of toast, everyone is at it—even scientists. One immediately brings to mind the story of the Italian astronomer Schiaparelli, who unwittingly joined the barely visible dots of craters on distant Mars into "channels" or, in Italian, *canali*—an illusion (compounded by mistranslation) that led American astronomer Percival Lowell to posit the existence of a globe-spanning system of canals, moving quantities of water from the martian poles, dug by a technologically advanced civilization under threat of extinction by drought. From this very human error comes H. G. Wells's stirring tale of Earth's invasion by Mars in *The War of the Worlds*; Orson Welles's notorious radio adaptation that had terrified crowds flocking into the streets, watching the sky; Edgar Rice Burroughs's Barsoom fantasies; and much else. Even today, when we know perfectly well that John Carter of Mars is a character in a pulp fantasy, and that tripods squirting death rays are unlikely to be found in New Jersey, nor, as it may be, suburban Surrey, scientists with their intellects vast, overheated, and oversympathetic look at Mars with eyes perhaps overwelcoming of the possibility of life.

All of which areological digression leads me very conveniently to another remarkable ability of humans—that of telling stories. Chains of unconnected craters became lines which became canals which became, implicitly, a heroic narrative of a great civilization struggling against extinction—and, more explicitly, thrilling yarns of interplanetary warfare and high adventure.

So, not only are we good at spotting patterns, even if nonexistent ones, we tend to weave them into tales as ways of making sense of what might otherwise be sets of disconnected and therefore worrying phenomena. This ability is so ingrained that it even haunts our subconscious. Things that go bump in the night are seamlessly woven into the stories we tell ourselves in dreams. It is easy to see how our ancestors, living much closer to nature, the unknown, and the reality of sudden and unexplained phenomena than we do nowadays, would hear thunder in the mountains and console themselves with stories of angry gods. And because telling stories is what we do, even without conscious intervention, it's easy to underestimate how the power of narrative undermines our efforts to make sense of the past, in any clear, cool, or rational way.

Fossils present direct evidence of the prehistoric past and for evolutionary change, but they are very thin gruel on which to build a narrative—rather like the dots of individual craters that Schiaparelli and Lowell willed into line segments, canals, and stories. This has not stopped people doing that very thing, but if they do, they must be aware that such a narrative is very likely to be colored as much by past prejudices as by present evidence. Fossils don't tell stories. *We* tell stories.

And so, in popular science books, particularly older ones, you'll hear tales of the ages of life—the Age of Fishes, the Age of Amphibians, the Age of Reptiles, the Age of Mammals, just as if they were biblical dynasties, one succeeding the other, replacing the one before—inevitable, seamless, majestic, culminating in the Age of Man.

One of my favorite examples of this tendency is indeed called *The Age of Reptiles*. This is a marvelous 110-foot mural at the Peabody Museum of Natural History at Yale, painted in 1947 by Rudolph F. Zallinger. It depicts almost 350 million years of prehistory as a landscape, with time moving from left to right. This image—and the various dinosaurs and other animals pictured within it—has become iconic.

Even before I knew what a dinosaur was, I knew of Zallinger's work. For my fifth birthday I was given a book called *Wonders of Nature*, whose back cover was adorned with a kind of condensed version of *The Age of Reptiles*. I still have the book (I'm looking at it as I write this), and there is no indication of what the animals on the back cover were—no caption, no acknowledgment, no nothing. But I was captivated, nonetheless.

A little later when I was at school, I loved the Life Nature Library, a series of color primers on science brought out by Time-Life. It was in the book called *Early Man* (1965) that I first came across the now canonical image of human ancestry, depicted as a left-to-right "march" of ancestry and descent, the conceptual progenitor of my figures 1 and 2 earlier in this book, and of many others. The picture was called *The March of Progress* and was painted—hey, you're way ahead of me—by Rudolph Zallinger. I do not think Zallinger had any intention to mislead. He did not, after all, draw arrows between the various reconstructed hominins. But his images have the power they do because they trigger our innate desire for narrative. Once we see them, we cannot help but put arrows between the images and think of them as ancestors and descendants.

In this book, therefore, I have as much hope of curing you of a perfectly natural desire to make stories out of disconnected dots as persuading the tide to turn at my command. All I can do is show you how very hard it is, in reality, to justify evolutionary narratives created from fossil evidence; invite you to wonder why it is that you create the stories you do; ask you to inquire how your status as a human being colors your view (quite naturally) of what you think ought to have happened; and, once that has been accounted for, imagine what other stories might be possible given the evidence at hand.

What would our picture of human evolution be like had we evidence of many more kinds of fossil hominin living into the recent past, or fewer; or had we persuasive evidence of nonhuman hominins living on this planet today?

The irony is that—I guess—our picture of evolution would be very similar to the one we have now. Such are our prejudices about progress; such is our overwhelming need to tell stories, that we'd have spun a tale of upward progress and improvement, culminating in Man, no matter whether we had ten times the fossils we have now, or none, and irrespective of the provenance or the poverty of the ingredients.

At this point I should add a few cautionary paragraphs. I made similar points in my book *In Search of Deep Time* (1999), but my words continue to be misconstrued more than a decade later—quite willfully and deliberately, and with intention to deceive. The culprits have been creationists, who quote extensively from *In Search of Deep Time* in support of their view that evolution is somehow "wrong," such that even a "prominent evolutionary biologist" such as myself "admits" this. Despite repeated attempts to expose creationists for such context-free quote-mining, the creationists are still at it.

Perhaps the most shameful activity in which creationists indulge is to present a distorted version of science to parishioners who might not know any better. A few years ago an elderly neighbor came up to Mrs. Gee in the street and gave her a pamphlet that she thought might be interesting, as it mentioned me. I sighed—it was Christian literature in which my various utterances on evolution had been quote-mined in support of creationism. Readers in the United States, who are more used to this sort of thing, will be either comforted or disturbed to learn that creationism runs deep in mainstream English churchgoers, not to mention synagoguegoers and mosquegoers.

It is quite true that I have said quite a few grandstanding things about evolution, and if taken out of context, you can see why they fill creationists with glee. Here is a choice selection from *In Search of Deep Time*, extracted and presented out of context by a creationist website,[1] with responses by me.

> The intervals of time that separate fossils are so huge that we cannot say anything definite about their possible connection through ancestry and descent.

Leaving aside the assertion by some creationists that no such intervals of time exist, the creationist spin is that no connection exists between ancestors and descendants, because of the unsupported presumption that God made everything separately. The proper answer (made clear elsewhere) is that ancestors and descendants exist—the community of all life is evidence for this—but we could never know that any fossil we find is an ancestor or descendant or anything else. Quite apart from anything else, the concept of Darwinian evolution is more elegant as a theory than anything offered by creationism, because it explains the community of all life without recourse to any other factors, whether they are Lamarckian *besoin*, Goethean cosmic strivings, or God.

> New fossil discoveries are fitted into this preexisting story. We call these new discoveries "missing links," as if the chain of ancestry and descent were a real object for our contemplation, and not what it really is: a completely human invention created after the fact, shaped to accord with human prejudices. In reality, the physical record of human evolution is more modest. Each fossil represents an isolated point, with no knowable connection to any other given fossil, and all float around in an overwhelming sea of gaps.
>
> Dinosaurs are fossils, and, like all fossils, they are isolated tableaux illuminating the measureless corridor of Deep Time. To recall what I said in chapter 1, no fossil is buried with its birth certificate. That, and the scarcity of fossils, means that it is effectively impossible to link fossils into chains of cause and effect in any valid way, whether we are talking about the extinction of the dinosaurs, or chains of ancestry and descent. Everything we think we know about the causal relations of events in Deep Time has been invented by us, after the fact.
>
> To take a line of fossils and claim that they represent a lineage is not a scientific hypothesis that can be tested, but an assertion that carries

> the same validity as a bedtime story—amusing, perhaps even instructive, but not scientific.

The chain of ancestry and descent we construct after the fact is just that—a human construction, a way of interpreting the evidence. However, this does not negate the existence of evolutionary ancestry and descent. I suspect creationists are sometimes motivated by the suggestion that when evolutionary biologists are in company, away from the public eye, they "admit" that evolution is wrong, while perpetuating some enormous cover-up to set before the masses. One shouldn't like to say in print that this is paranoid, but any suspicion of such a cover-up is immediately scotched by the fact that many books making these points are widely available to the public. *In Search of Deep Time* was hardly a massive best seller—but it wasn't some dark secret either, the existence of which could only be vouchsafed to the Elect.

> All the evidence for the hominid lineage between about 10 and 5 million years ago—several thousand generations of living creatures—can be fitted into a small box.

To which we say—so what? This illustrates the extreme poverty of the fossil record, offering a caution to anyone who would use this evidence on which to base an evolutionary scenario. It doesn't dent evolution in any way.

Creationists quote material out of context to give you the misleading impression that anyone has any doubt about evolution's status as a theory so well worn that it can be accepted as true. I hold that view now, just as I did more than ten years ago when I wrote *In Search of Deep Time*—and in that book, too, I made my views clear, except that creationists have chosen not to mention them. "If it is fair to assume that all life on Earth shares a common evolutionary origin," I wrote on page 5, going on to make clear that this is my assumption throughout the book. Creationists are very good at either ignoring such statements—or, if they mention them, say words to the effect that if even "prominent evolutionists" who explicitly sign up to the fact of evolution can produce statements in which evolution is doubted, there shouldn't be any reason for anyone else to "believe" in evolution, either. And they just keep rolling along: enter "Henry Gee" and "creationism" into a search engine of your choice, and they'll be all over you like an embarrassing rash. That said, I have had some robust and heartening support.[2]

The sad thing is that no matter how hard I fight, the creationists will still take quotes out of context, because that's the way they do what they call "science." Like all pseudoscientists and peddlers of charlatanry, they don't investigate anything systematically. They just pick out the things they like and discard anything else, even flat statements to the contrary. Now, I could try quoting scripture out of context to show how such a procedure can be used to mislead. For example, "There is no God."[3] But that approach might be too subtle. That said, I refuse to modify my thoughts for fear of being quote-mined by idiots. I tend to regard creationists as an occupational hazard, rather in the same way that those who go walking in the dark, looking up at the stars, will occasionally tread in a pile of dog shit.

In the end, one can only feel a kind of pity for creationists. Many believe in the literal truth of Genesis, despite the fact that the Bible was written at various times by different hands, and despite the fact that the text has been translated into English from classical Hebrew, a language so tricky that people of formidable learning, such as St. Jerome, Thomas Aquinas, Rashi, and Maimonides, spent their lives trying to understand its nuances in order to extract meaning from the same scriptures to which many people of perhaps lesser intellects cleave without question. Those living in medieval times had perhaps no good reason to doubt the literal truth of the Bible. People living today do not have this excuse.

Evolution itself, however, is *not* in question.

Evolution happened, and there is, out there, a true skein of ancestry and descent between some primordial blob and every creature living or extinct, but we can never trace it with absolute certainty, or if we stumble across part of it, we can never know that we have done so.

What is in question, however, are the ways we interpret the evidence given to us by fossils. It's not that fossils don't provide us with primary evidence for evolution as a fact, because they plainly do so. What is at stake is a common misreading of evolution that flatters our prejudices: that we are the pinnacle of creation, and the various stages toward this manifest destiny can and should be discernible in the fossil record. The picture of a simple, linear evolution, with each species of human being succeeded by a more sophisticated form, "culminating in Man," can only be extremely inaccurate, and also misleading.

Although it's fun to take potshots at creationist misbehavior, it is perhaps worth asking why creationists remain indefatigable despite the evidence, devoting such time and effort and skill to monitoring the writings of "evolutionists" and extracting such morsels that suit them. When you take a step back, you can see that we have seen this mindset elsewhere, among those scientists who look into the unknown and see a set of circumstances that dashes the more comforting scenarios on which they have perhaps based their reputations. I think that what motivates creationists and such scientists is a very human fear of the unknown, and the uncertainty that accompanies it.

The fundamental difference between religion and science is that the former is all about the celebration of certainty, whereas the latter is all about the quantification of doubt. Creationists understand this instinctively. What they cannot afford to see happen is that people start wondering about their place in the universe, and asking whether the certainties in which they have been raised might not be so certain after all. They are so desperate to avoid this that they have tried to subvert science by invoking a bogus replacement called "creation science," perhaps the most shocking oxymoron ever invented, given that creationism and science concern such fundamentally different things.

Scientists have been less ready to appreciate this distinction, to their cost. When confronted by creationists, they are inclined to close ranks and present a united front of "fact" against "mythology." Such a strategy only plays into the creationists' hands, leaving them free to mine the works of evolutionary biologists for quotes—the subtext being that scientists are always squabbling behind the scenes, and the united front they want you good honest folks to believe is a cover-up.

In my view the best way that scientists can confront creationism is to be as honest as possible. Science is not all about truth given to us by authority, but doubts that arise from the ground up. You, the citizen, should never be afraid to ask a silly question—and you, the scientist, should never be afraid to admit that you don't know the answers.

So much for the fossils. What of humans living today? What actually defines a human being, so that you can tell one apart from, say, a postbox, or the back end of a cow? What is Man, if no longer the microcosm that measures the macrocosm? To quote scripture again, "What is man, that thou art mindful of him?" (Psalms 8:4).

My task in the rest of this book is to show that this question is meaningless. Were one to accept the argument I put forward in chapters 2–5 of this book, the very idea of a distinctive nature of humanity is already questionable. Once one acknowledges that the ladder of creation with *Homo sapiens* at the top is the result of a fundamental misreading of evolution, you can see that when viewed objectively, we humans are no more or less deserving of special consideration than any other species. There is certainly nothing so special about humanness—as opposed to hamsterness or geraniumness—that demands the elevation of humans to a higher order of being.

Those of a certain turn of mind or upbringing will no doubt balk at this, saying that humans are different from other animals (and plants, and bacteria, and fungi, and so on) because they have an immortal soul. It's hard to argue against convictions founded on belief rather than empirical evidence, except to counter that each and every species has attributes that allow us to recognize it as such. The Madagascar star orchid, for example, is recognizable by the extraordinarily long floral spurs of its flowers, penetrable only by the very long tongues of a particular species of moth. Such features can be identified and quantified, which cannot be done for the soul, begging the question of the existence of such an attribute.

Even if we leave such imponderables as the existence of the soul to theologians and philosophers, we run into another problem: it's very hard to define what we think is special about humanity because it's we, the humans, who are composing the definitions. Objectivity is impossible. The validity as such of any we recognize in ourselves is compromised by an unavoidable subjectivity. Were we all Madagascar star orchids, we would no doubt measure our exalted state by the lengths of our floral spurs relative to those of other orchids.

In the rest of this book, I take a brief tour of several attributes that at some time or another have been regarded as unique to humans. These include bipedality, technology, intelligence, language, and finally sentience or self-awareness. It turns out that most if not all have been seen in one or more nonhuman species—or once one has accounted for a human bias in investigating such attributes, they turn out to be no more special than any other feature of any other organism.

The order in which I examine these attributes is not random, but dictated by how easily we can find actual biological evidence for the evolution of these traits.

Bipedality, for example, can be assessed directly, by looking at fossil bones. We can judge by direct inspection whether a given fossil creature habitually walked on its hind legs, or not. So much so that bipedality is seen as a hallmark of the hominins. A fossil ape is marked as belonging to the hominins if it is bipedal, almost without reference to any other feature. From this is would be easy to imbue the acquisition of bipedality as something special, the first step (pun intended) in the inevitable journey to the human state, as if technology, language, intelligence, and so on would surely follow. Bipedality, however, is just one peculiar posture adopted in a group of animals in which the adoption of peculiar postures is commonplace. Human bipedality is a posture seen nowhere else—but one could say the same for knuckle walking in chimps and gorillas, brachiation in gibbons, and the four-handed swing of orangutans. Furthermore, there have been one or two fossil apes, unrelated to hominins, that were more or less bipedal, and their fate was extinction without achieving technology, language, and so on (as far as we know).

Technology, too, leaves traces in the historical record, although—as we have seen—it is not always easy to link a tool with its maker. When the first stone tools were discovered at Olduvai Gorge alongside the remains of fossil hominins such as *Homo habilis*, people tended to associate the fact of toolmaking with increased intelligence, in particular an attribute known as "planning depth." To make a tool, a creature should have some "idea" of what the result should look like, or be used for, and therefore have some notion of the future and its place within it. Subsequent work has questioned this idea in two ways.

First, other animals are known to make and use tools. Tool use has been seen in apes, various birds, even octopi, in the sense that an animal will use some object to help it achieve some goal that it would not manage unaided. In some cases the object has even been modified for use—an important distinction. This questions the idea that tool use is a distinctively human attribute.

Second, if tools—or technology—can be defined loosely as devices created by the modification of materials to achieve some specific end, then many organisms have produced technology that makes the earliest stone tools look puny indeed. One might include termite mounds, the nests of bowerbirds, bacterial stromatolites, or even the Great Barrier Reef, as examples of technology. One could always object by saying that such structures are not technological because they were not made using "planning depth," but such objections run into problems of subjectivity. How can one "know" what a New Caledonian crow is "thinking" while it fashions a piece of leaf into a probe? Is it thinking about how it will use the tool it is making? Is it thinking about something else entirely? Is it thinking at all?

One does not, in fact, have to inquire as to the thoughts of crows, because there are good arguments for saying that the earliest stone tools, beautiful though they are, required as much planning depth to produce as the nest of a bowerbird—in which case one can say that stone tools are no more "special" than any other structure created by living organisms. The alternative is outrageous—that organisms as "lowly" as bacteria, coral polyps, or bacteria have "planning depth."

There is a third possibility, however, which is that notions such as "planning depth" are entirely illusory and products of the view of human evolution that is narrative and linear. In the real world, organisms just do what they do because that's what they need to get by on Darwin's tangled bank. Bees make beautiful honeycombs, coral polyps make mighty reefs, and humans make shoes, ships, and sealing wax, and one need not inquire as to their internal motivations, if any, to assess the adaptive value of these attributes to the organisms concerned.

Technology is usually seen as a hallmark of intelligence, but once one acknowledges that the link between the two is tenuous at best, one starts to wonder what intelligence is, such that it constitutes a unique attribute of humans. There are perhaps few points of discussion more emotive than the meaning of "intelligence." Like the mythical city that is forever on the horizon but that can never be reached, the meaning of intelligence has forever remained beyond our grasp. How is it defined? How is it measured? Do any measurements (such as "intelligence quotient" or IQ) mean anything apart from the ability to do IQ tests? Will such tests only ever be able to assess aspects of that thing we call intelligence, rather than intelligence itself? In which case, can intelligence be thought of as a discrete, unified attribute, rather than a set of attributes unified after the fact? Is intelligence—whatever it is—heritable?

Intelligence is something like jazz—you know it when you find it, but it's almost impossible to define. And if measuring intelligence in humans is difficult, measuring it in other species is probably impossible. You might regard someone who can solve the *Times* crossword in less than twelve minutes as intelligent—but this ability might say as much about a person's upbringing or cultural milieu as any innate capacity. No crow or dolphin or octopus—all animals commonly regarded as intelligent—has ever been caught even attempting the *Times* crossword.

Scientists have sought to understand the evolution of intelligence by fairly crude measures such as brain size or brain volume relative to body mass, but this idea soon runs into problems. *Homo floresiensis* had a very tiny brain indeed, but appears to have made tools (not that this need say much about intelligence, as I noted above). Other animals regarded as intelligent, such as crows, have higher brain volumes relative to body mass (the "encephalization quotient" or EQ) than other birds seen as less intelligent—but their brains are tiny in absolute terms, are as capable in many respects as those of humans, and are in any case constructed entirely differently. All of which leaves any simple equation of brain size with intelligence gasping in the dirt.

Language leaves no fossil record. But when we listen to the kind of language we use every day, we can't see much distinction between the messages that language conveys and the messages that animals exchange, even though they appear mute—messages about social and sexual status. Anthropologist Robin Dunbar thinks that language originated as a form of social grooming, perhaps no different, qualitatively, than baboons picking lice off one another's fur. It is perhaps no coincidence that people conventionally greet one another with inquiries as to their state of health—one might say the same of dogs who, on meeting, sniff one another's bottoms. Human language is special only in its peculiar mode of delivery, not in its function. It is also probably no coincidence that no human group so far discovered is without language, so it requires neither special skill nor intelligence to master.

Sentience is perhaps the knottiest problem of all, because we have to be self-aware to discuss it—or do we? I shall propose, perhaps surprisingly, that sentience is a phenomenon that we experience relatively rarely, if at all, and can often be regarded as a syndrome of teenagers and young adults whose brains are in the throes of development. On the contrary, twenty-four-hour sentience would be a debilitating handicap rather than an evolutionary advantage. Moreover, recent work shows that at least some nonhuman animals, crows in particular, are capable of behavior that we might regard as sentient, suggesting that self-awareness is not an attribute unique to humans or even mammals.

On the other hand, our perception of sentience might itself rest on a grave error. As Daniel Dennett describes in *Consciousness Explained*, it depends on the ability to imagine ourselves as participants in the drama of our own lives, which depends on a conceit called the "Cartesian theater," which is itself flawed. If this flawed model of sentience applies to the way in which we think other sentient animals think, then sentience is a red herring that applies as much to other animals as it does to humans. We see it in animals because that's what we see in ourselves.

My aim in the rest of the book, therefore, will be to show that our view of these attributes as uniquely or specially human is an illusion, created by our view that evolution is linear and progressive—as John Zachary Young put it, "culminating in Man."

I have a suspicion that the distinction between humanity and the rest of creation is a relatively recent phenomenon. Folk wisdom and popular mythology—as opposed to state-sanctioned or official religion—have always respected nonhuman creatures, and even inanimate objects, as individually powerful with distinctive attributes that might be the envy of humans. Animals in folklore and nursery tales are able to converse with one another, and even humans, using human language. Such animals are intelligent, with motivations as complex as those of any human.

In her book *The Animal Connection*, paleontologist Pat Shipman contends that one of the things that make us human is a deep connection with animals. When human beings were first able to paint pictures, they painted pictures of animals. Everyone is familiar with the remarkable cave paintings—even if as reproductions—from caves such as Lascaux in France and Altamira in Spain, as well as from a host of other sites. But it had never occurred to me, until I read Shipman's book, to wonder why the pictures are almost exclusively of animals. There are no portraits of human beings—certainly nothing to rival the naturalistic accuracy of the animal paintings. People appear as handprints, cartoonish stick figures, or grotesque "Venus" figurines. Neither are there pictures of plants, landscapes (plains, mountains, volcanoes), or weather (sunshine, clouds, rainfall, lightning), which one would have thought would have figured large in the lives of the artists. I suspect that the painters did not see themselves as we are inclined to do—as somehow separate from the animal world—but very much a part of it. Animals were to be hunted, to be sure, but also to be venerated.

Importantly, people were accustomed to seeing the world from the animals' point of view. This can be seen in very early depictions of people dressed as animals. "Therianthropes"—sculptures of people with the heads of animals—are among the earliest known human artworks.[4]

This tendency goes right down to the present day. When I was a student I used to play piano accordion with a band that accompanied a side (troupe) of Morris dancers. For those who don't know, Morris dancing is a type of folk dance typical of southern England. Modern metropolitan types tend to dismiss the sight of grown men (Morris sides are traditionally all-male) dancing around with ribbons and bells attached to their ankles as twee, even effeminate. The reality is rather different. Morris men, thundering around a pub car park, are formidable, primeval, perhaps rather frightening.

One of the dances performed by my Morris side was called "Shooting the Badger." One of the dancers wore a badger mask, for all the world like an Ice Age therianthrope. The other dancers carried short staves and circled round, beating the staves together in rhythm, until—at a signal—they all pointed their staves inward at the badger and said "Bang!," at which point the badger fell down "dead." It was the task of the lead dancer—the "squire"—to accost an attractive young woman from the audience to "revive" the badger—after which the badger got up and the dance continued. I do not know this for sure (I have never investigated it), but this dance looks very much as if it harks back to fertility rituals from the earliest days of modern humanity in Europe, as represented in early cave art, in which animals are objects of awe and reverence rather than subservience.

Until recently, and even today in some societies, animals carry a social or even a legal status equivalent to that of humans, so much so that they can even be tried in a court of law.[5] During the Napoleonic Wars, a French ship was wrecked off the northeast coast of England, and the only crew member that made it alive to the nearby port of Hartlepool was a pet monkey. The citizens of Hartlepool—never having seen a Frenchman before and not wishing to take any chances—put the monkey on trial as if it were a French serviceman and sentenced it to death by hanging. To this day, citizens of Hartlepool are sometimes known as "monkey hangers"—an epithet that Hartlepudlians wear with pride.[6] In terms of the folk wisdom of the age, the citizens of Hartlepool were acting entirely logically—provided that one's worldview was more accommodating of the notion that human beings and animals were of equivalent status. It could well be that the distinction between humans and animals, something we very much take for granted as having pertained since time immemorial, is in fact a more recent, post-Enlightenment fancy.

7: *The Way We Walk*

It happened a long time ago, but the experience was so traumatic that I remember it as if it were yesterday—the moment when the outraged, elderly professor pinned me against a wall and harangued me for having rejected his paper on why human beings got up on their hind legs and walked. Human beings became bipeds, yelled the prof, to free the hands so that mothers could cuddle infants close to their chests. How could I have had the temerity, screamed the empurpled sage, to have rejected a paper that made so much sense?

One of the problems with human evolution, as opposed to, say, rocket science, is that everybody feels that their opinion has value irrespective of their prior knowledge (the outraged academic in the encounter above was a scientist, but not a biologist, still less an evolutionary biologist). It's obvious to see why—we are all human beings, and we are all bipeds, so we think we know all about it, intuitively. What we think about bipedality "stands to reason." Now, I'd be the last to disparage anyone who wanted to express an opinion, however cockeyed, but it is sometimes the case that the most perplexing problems are those that seem the simplest at first sight.

It is always a wonder to me that there is still much to be discovered about something so screamingly obvious as the way we humans walk. However, much about human walking remains to be understood. Why, for example, do we walk the way we do? Why, when moving faster than a certain speed, do we start to run? Why do we walk upright at all, when other animals get by perfectly well on all fours? These and other such questions are still being debated by scientists. I remember publishing a research paper showing that there was a perfectly feasible gait, somewhere between walking and running, which people never used.[1] I enjoyed demonstrating the gait to my colleagues, as if it were something out of the famous Monty Python sketch about the Ministry of Silly

Walks. We might not know the arcane secrets of the universe, but we are all perfectly familiar with walking and running, so how could there be a third, distinct gait, available all the time for our use, and we somehow missed it?

As anyone who has watched a cruising toddler will attest, simply the act of standing up on two feet requires a great degree of control, and scientists still have a great deal to learn about how this is achieved—and this is with modern human subjects who can be watched and their activities measured. And even after all this, robots that can walk with anything like the natural grace of a human have yet to be built. How much harder it is to learn about how bipedality evolved, still less why.[2] The very fact of bipedality remains a taxing problem for those versed in fields as diverse as evolutionary biology, mechanical engineering, and robotics. It's not the easy problem that people so often imagine.

The common or garden explanations put forward by armchair theorists tend to avoid the problems that engage serious scientists—problems of energetics, and posture, and balance, and anatomy, and neuromuscular control, in other words anything that might require some actual scientific training and a facility with at least the basics of mechanics—and cut to the chase of why the ancestors of humans became bipedal. These explanations are invariably teleological. That is, they are driven by some inherent purpose or striving, in the manner of Lamarck—or, indeed, of the popular model of evolution as "progressive," which I have demonstrated as erroneous. For example, humans got up *so that they could* free their hands in order to make tools or grasp low-hanging fruit;[3] or *in order to* cuddle babies close to their chest; or *in order to* see longer distances; or *in order to* live better in open country rather than in forests, as our ape cousins still do.

All such arguments are easily demolished. For example, many animals make tools, irrespective of whether they have hands; nonhuman animals of all sorts have no problem cosseting their young close to their chests; many animals are tall, or can make themselves so, without being bipeds; many large primates such as baboons live in open country and do so on all fours without extravagant distress. So why should bipedality be in any way remarkable, a qualitative advance over what other animals can achieve? Why not stay on four legs and evolve longer legs? Or longer necks? Why not evolve jumping or hopping?

Another idea is that bipedality evolved *in order to* make it easier for people to keep cool in hot climates.[4] A biped presents a much smaller cross section to the sun—just the top of the head rather than the whole body. The rest of the body, not pointing sunward, is thus free to radiate away any excess heat. This idea makes sense, in part, because it seeks to explain a suite of other features of humans that don't immediately seem connected with bipedality. These include our hairlessness relative to other primates, and the presence of large numbers of sweat glands in our skin. Taken together, you can see how a creature with exposed skin and plenty of sweat glands could have stood up in a breeze to cool off—an advantage in the hot, dry climates of Africa in which the human lineage is thought to have evolved.

This all seems fine, except that there are lots of other animals that live in hot climates that are both quadrupedal and very furry. And the idea also doesn't explain another feature of human heads, namely male-pattern baldness. Why should males become bald, exposing their scalps to the direct glare of the sun, while females generally retain their heads of hair?

The problem is that you can come up with any number of other ideas that "explain" any suite of features you choose, all of which have much to recommend them, and none of which can be shown to have any more scientific validity than any other. Just come up with a scenario, and then cherry-pick the features of modern humans you need to make the theory work, and ignore any others.

An example of this kind of approach is the "aquatic ape" theory, promoted for many years by Elaine Morgan.[5] Morgan selects a range of features of modern human physiology and behavior to suggest that there was once a period in human evolution during which humans were aquatic—that is, lived in and around water, and became adapted to an aquatic environment in a way that our close ape cousins did not. This idea is perhaps the most developed of all the various ideas I have described as teleological, and the subject of several books that have gained a degree of respectable support.

The anatomy of humans is certainly peculiar in many ways relative to that of apes such as chimpanzees and gorillas. Humans are much fattier than apes and are much less hairy. In contrast to almost all other primates, humans are capable swimmers, and newborn babies appear to have an inborn capacity for swimming. In these respects humans are less like apes than, say, seals and other aquatic mammals, which are relatively fatty and hairless, compared with their purely terrestrial cousins. Humans, in contrast with apes, have historically eaten a lot of seafood, a diet that offers minerals and fatty acids essential for development, especially for the brain and nervous system, but otherwise hard to come by unless humans once spent a great deal of time in and around water.

There is a lot more to the "aquatic ape" idea than that, of course, but from this brief description you can, I expect, already identify some flaws. The first is that it's always a problem identifying features that humans have now and inferring that they must have had some adaptive value in the past. It's entirely true that humans seem to have an unusual fondness for seafood—but we still do, and it remains an important part of our diet. But we humans also consume an extraordinary range of foodstuffs compared with other animals, including substances that are noxious or even bad for us, such as capsaicin (the substance that makes chili peppers hot), alcohol, tobacco, and dangerous drugs. What of body fat and hairlessness? If they were once selectively advantageous for water-loving humans, why are we still relatively fatty and less hairy than other apes? Presumably other factors have come into play that might have nothing in particular to do with an aquatic stage in our history.

Second, it's notoriously hard to infer habits from anatomical structure. If a busload of Martian anatomists came across the skeleton of a goat, the one who said that goats would be good at climbing trees would be laughed off the planet. With their long, spindly legs and complete lack of grasping hands or feet or tail, you'd think that goats could climb trees as handily as giraffes can ride unicycles. However, it so happens that goats are surprisingly good at climbing trees.[6] And you'd never guess from her fur that my golden retriever, Heidi, is a capable and strong swimmer, regularly braving pounding North Sea surf to retrieve a stick or ball. Heidi is very far from hairless—indeed, she can brave the cold and wet of the sea because she has more fur, rather than less, an extra layer of underfur that keeps the cold and wet away from her skin.

The relative hairlessness of humans is complicated, however, by sex. It's easy to say that being relatively fatty and less hairy might be a sign of aquatic ancestry, but this doesn't explain why human females are very much fattier and less hairy than human males. In addition, men become more hirsute as they mature (excepting male-pattern baldness, as I mentioned above). Sex differences in fat content and hairlessness are intriguing and demand explanation. Their presence might even shed light on why we humans are bipeds.

Jared Diamond suggests that relative hairlessness, combined with differences in fat distribution, might be connected with what Darwin called sexual selection.[7] This is the tendency for the different sexes to exhibit their own traits that they use to attract the attentions of the opposite sex. The most famous example is the peacock with his extravagant tail, which he displays to attract the attentions of the much less extravagantly endowed peahens. Sexual selection arises because males and females contribute different amounts to the next generation. Typically, a male will contribute sperm, which are easy and cheap to make in large quantities, and will seek to inseminate as many females as possible. Females, on the other hand, contribute eggs, which are expensive to make and rare, and so will have much more at stake. This is why males are showy, and females are choosy—females make a much greater investment in the next generation, so have more to lose if this investment isn't recouped in terms of numbers of strong, healthy offspring. It is incumbent on males, therefore, to demonstrate to females that they would be appropriate mates, usually by some proxy such as a display (showy plumage, mating calls, and so on) that illustrate their suitability.

Sexual selection is a vibrant subject of study in modern evolutionary biology.[8] Evolutionary biologists are still arguing about what it is, precisely, that choosy females are selecting in prospective mates. We know that birds (say) suffering from parasitism or disease look drabber and more droopy than those in the peak of health.[9] Is showy plumage therefore a reliable mark of a healthy genetic constitution? Does a sports car, an indicator of high status or a fat bank balance, mark a man as a better potential mate than had he been seen riding a rusty bicycle? This is the "good genes" idea—females choose males based on signs of good general health.

Or is the association of a showy male trait somehow linked—perhaps by chance, to begin with—with female choosiness for that particular trait, reinforcing one another down the generations? This is the "runaway process" first elaborated by the brilliant geneticist and statistician Ronald A. Fisher.[10] To put it another way—need there be any particular reason why red sports cars are sexier than (say) blue ones? Peacock tails confer no obvious advantage on a peacock apart from attracting peahens, but why the elaborate tail rather than (say) a mating call, as in nightingales; or the construction of a bower, as in bowerbirds? The answer could be that the male display trait, perhaps entirely random to begin with, has become linked, genetically, with the female preference for that trait. Successful partners have male offspring that display the trait strongly, and daughters that are strongly attracted to that trait, so that the two traits have become reinforced down the generations. Initially, there is no reason why the selected trait has any selective advantage at all, and its choice might be completely fortuitous. If sports cars, why *red* sports cars?

There is another aspect to this, too, related to self-advertisement. Extreme traits such as the long tails of peacocks are actually a disadvantage. They are expensive to make and maintain in terms of resources, and interfere with important aspects of daily life—such as the ability of the peacock to fly away from predators. Such displays seem to say that the male is not only fit to be a mate, but so fit that he can survive perfectly well despite having to support such seemingly profligate habits, as if he has fitness to burn.[11] Sports cars are expensive not just to buy but to maintain, as anyone attempting to buy an insurance policy for one will attest.[12]

Irrespective of its internal mechanics, nobody doubts that sexual selection exists. Here, by way of making my own contribution to the shuddering pile of teleological arguments that purport to explain why humans are bipedal, I'd like to suggest how sexual selection might have played its part. I'm not suggesting it's correct, or even that it's original.[13]

Still less would I pin anyone against a wall and shout at them about it.

If standing upright does one thing, it exposes one's breasts or genitalia to full view—especially if one has relatively little fur, and no clothes. No other ape is as habitually bipedal or as hairless as humans, and these features might be connected with another human peculiarity, that human females do not show any physical sign of when they are in estrus[14]—"in heat"—that is, sexually receptive such that sex has a high probability of producing offspring. Other apes are not only hairy and quadrupedal, but their females make it perfectly obvious when they are in heat, by displaying large swellings in the genital area. The breasts of ape females are also tiny, covered with fur, and swell only when they are pregnant or lactating. When apes mate, they do so in full view of other apes.

Estrus in human females is concealed, even from the female herself. No external sign betrays when a human female is more or less likely to conceive. In addition, the breasts of human females are more or less prominent at all times, and the fact of hairlessness makes them more prominent still. Being bipedal makes breasts more obvious even as estrus is concealed.

It remains a mystery why estrus is concealed in humans[15]—just as it is not obvious why humans tend to have sex in private. The usual explanations concern the tendency in humans to be monogamous and form stable pair bonds, but this argument has its own problems. Human societies are particularly variable as regards their mating systems—polygamy is widespread—and even when societies are nominally monogamous, both males and females cheat on their partners (what scientists call "extra-pair copulations") more frequently than polite society admits.[16]

In these respects—cheating, and having sex in private—human sexual habits have more in common with nesting birds than with apes. Much work on nesting birds reveals multigenerational family structures that are much more complex than anything seen in apes, but highly reminiscent of human proclivities, including the tendency to overt monogamy and covert extra-pair copulation[17]—which by definition happens in private. Like birds (but unlike apes), humans are prone to elaborate sexual display by males, with consequent choosiness by females, and evidence has also come to light that in birds, females use their own appearance and behavior not just to attract males, but to compete with other females[18]—another notable human trait that I shall discuss again later. All this aside, it seems possible that bipedality is related to hairlessness, sexual display, and the still unsolved problem of the concealment of estrus in females.

How is this related to sexual selection? Let's look more closely at the secondary sexual characteristics of humans inasmuch as they relate to body fat and hairlessness.[19] As any middle-aged male reader will know, males are in general rather lean, and if fat starts to accumulate, it is around the gut and internal organs, and then generally after a male has done all his reproducing. Females, though, even when young, are much fattier than males. The percentage of body mass that is fat is 10.4 points greater in females than in males of the same body mass index.[20] Body fat in women is spread all over, just under the skin, and the skins of females are on average smoother than the skins of males of similar ethnic background, a difference that might have been maintained by sexual selection.[21] Females also have skin that is significantly paler than that of males of similar ethnicity. Apart from that, fat deposits in females are concentrated in the upper arms, breasts, thighs, and buttocks. This difference in fat deposition leads to very obvious differences in appearance, and it seems likely that they have some connection with sexual selection. The historical male preference for plump, rounded women with ample *embonpoint* is proverbial, from the "Venus" figurines of the Paleolithic to the well-upholstered nudes of Titian, Rubens, and Renoir.

However, might it not be the case that standards of beauty are in part conditioned by culture, rather than purely by sexual selection? Contemporary "Western" standards of female attractiveness tend to emphasize a leaner physique, so does the conventional picture of amply bosomed women have more to do with changing cultural norms than a more general, more ingrained tendency? A recent study of Peruvian men unexposed to Western media showed that their idea of feminine attractiveness was strongly associated with fat. They preferred women with a pronounced "hourglass" figure, big busts and behinds.[22] Those men who had moved to urban centers, and who had been exposed to Western advertising, festooned as it is in slender models, tended to find slimmer women more attractive.

It's simple to find pat answers to such preferences. Historically, fatness in women has been associated with reproductive success. Women with more fat would have the nutritional reserves necessary to nurture a fetus to term, and nurse it afterward. In the past, and in traditional societies, to be thin was to be ill—suffering from some threatening disease such as tuberculosis, or laboring under a large parasitic load. It's easy to see why men have traditionally found fatter women attractive. Only today, when nutritional resources are more abundant and less episodic, is fatness seen as a disadvantage.

It might also be the case that men are looking for different things in women than women are looking for in one another. Competition between females over appearance has been documented in birds,[23] and in this context it is noteworthy that pictures of slim, attractive women are aimed not just at men (in pornography, for example) but at women: in women's magazines, advertisements for beauty products, fashion plates, and so on.

If it seems all too easy to find reasons why fat is attractive, it's harder to understand hairlessness, or, at least, patterns of hair distribution. If humans are generally hairless, they retain hair on their heads, and, when adult, under their armpits and around the genitals. Why? Armpit and genital hair makes sense in terms of devices to trap secretions meant to attract members of the opposite sex, or deter rivals—except that the role of pheromones in human beings is very much an open question.[24] Head hair is another problem entirely. In many cultures, luxuriant head hair is seen as attractive in women, whereas it is common for men to lose much of their head hair in adulthood. What is head hair *for*? There seems no good, adaptive reason for the presence of hair on the head (as opposed to anywhere else) than sexual selection, and this illustrates how secondary sexual characters need have no adaptive reason except that they are attractive to the opposite sex, very much in tune with Fisher's runaway process. This applies as much to the distribution of fat as hair. Consider—why do men find women with pale skin, luxuriant locks, and curvaceous figures attractive? One can come up with examples based on nutritional status, but only after the fact. There is no reason, a priori, why gentlemen don't prefer bald women with hairy ears and enormous feet.

If females standing upright expose their breasts to view, men standing upright expose their penises. The connection between bipedality and penis display seems less fraught than that between bipedality and the hidden estrus of females. Males are always sexually receptive—their penises do not lengthen and shorten with the seasons. The connection between penis display and sexual selection should be too obvious to underline. And it is a curious fact that the penis of the human male is much larger as a proportion of body mass than that of any other ape. This combination of unusually large size, open display, and relative lack of body hair seems to speak loudly of sexual selection as well as habitual bipedality. It is perhaps noteworthy that there are tribesmen in New Guinea in which the men are naked except for elaborate sheaths worn on the penis that emphasize their presence and exaggerate their size.

This topic touches on another distinctive feature of humans, which is clothing. Conventional explanations for clothing include protection against harsh environments, as well as compensation for lack of body hair (and the two might be related). Such explanations are, like conventional explanations for bipedality, prone to teleology. To be sure, few will find Inuit parkas, space suits, or protective goggles sexy,[25] but I suspect that clothing in general is as much about sexual display as anything more utilitarian. The penis sheaths of New Guinea tribesmen conceal as well as emphasize sexual features, just as much as the swimwear displayed by a glamour model, the basque of a burlesque diva, or the bustle and corsetry of a Victorian debutante. I suspect that the evolution and development of clothing is connected with sexual selection, the strange fact of the hidden human estrus, and, beyond that, bipedality.

If none of that convinces, try this. Sexual selection is distinct from natural selection because, in sexual selection, features can be emphasized, which in normal circumstances would be highly maladaptive. The tail of the peacock is just such a feature. Bipedality is another. Standing upright introduces a potential for all kinds of injurious stresses to the head, spine, and limbs that simply don't apply to quadrupeds. Back pain, related directly to bipedality, is a significant burden on the economy.[26] Bipedality becomes even more problematic for women during pregnancy, and the evolution of the particular kind of spinal curvature typical of humans can be related to the need for extra lumbar support during pregnancy.[27] To suppose that bipedality evolved for some reason or another is to belittle the immense changes in bodily form that the human frame underwent simply to stand upright as of habit, and the considerable disadvantages accrued in so doing. All parts of the body have been profoundly influenced by the acquisition of bipedality, even the head.[28] Only sexual selection has the power to generate something so maladaptive, so seemingly pointless, as a peacock's tail—or human bipedality.

Much of the foregoing is written at least in part in jest. I do not claim that bipedality evolved for the purpose of sexual display. The point I am trying to make is one that armchair theorists of bipedality fail to understand: that there can be no simple relationship between a proposed cause and a proposed effect. The consequence of one change has an impact on many other traits or adaptations, until the whole body is affected. In no trait does this seem truer than in bipedality. Bipedality means more than just standing on two legs. It requires the wholesale modification of the body, not all of it very effective.

But bipedality has evolved considerably since the first appearance of bipeds: it did not appear all at once. The awkward gait of the very primitive fossil hominin *Ardipithecus ramidus* (at 4.4 million years old, the earliest for which good skeletal evidence is known) shows that the first bipeds were not as refined as modern humans.[29] They could stand upright, they could walk, though not as upright as modern humans, but they probably could not run very well. However, footprints attributed to the fossil human *Australopithecus afarensis* (Lucy) from a million years later show that by this time, creatures close to the human lineage could walk just about as well as modern humans. Even so, the skeleton of this creature was still very different from modern humans: Lucy could walk, but her skeleton suggests that she might have been a better tree climber than modern humans are.[30]

The act of standing upright was followed, in sequence, by walking and then running—two gaits that demand very special, and rather different, adaptations. Daniel Lieberman and Dennis Bramble have recently proposed that many features of modern humans appear to be adaptations not to walking, as such, but to long-distance running.[31] These include a range of features throughout the body not directly connected with the legs and feet.

Here are just two examples. *Homo erectus* and modern humans have barrel-shaped rib cages, in contrast to the cone-shaped, wide-bellied rib cages of earlier hominins. This means that later hominins had "waists," which would have allowed the counterrotation of the arms relative to the legs while running. This is an extremely important aid to balance. Such counterrotation, however, would move the head from side to side with each stride, if it weren't for a corresponding reduction in the neck musculature to allow the head to be suspended independently. In human beings there is a ligament—the nuchal ligament—that connects the back of the skull with the back and shoulders. This allows the posture of the head to be maintained without effort. This ligament is not found in apes. It *is* found, however, in predators such as dogs, which track and hunt over long distances without tiring—just as traditional hunters do.

The current consensus is that bipedality was the first distinctive feature to have evolved in the human lineage, long before the expansion of the brain. Before many fossils had been discovered, of course, the view was that the large brain of humans evolved before the upright, bipedal stance: this conceit explains why the Piltdown forgery was so effective.

Bipedality might be a distinctively human feature—but is it "special"? Not really—apes have a variety of peculiar modes of locomotion, from quadrupedal knuckle walking (gorillas and chimps) to movement with all four limbs as hands (orangutan) or the forelimbs alone (gibbons). The evidence from *Ardipithecus ramidus* suggests that the distinctive modes of locomotion in each modern ape species are products of their own very special evolutionary circumstances, and not some relics of ancient times. Some extinct apes, not directly related to hominins, were even bipedal.

As a final note in this chapter, I refer you to the strange case of *Oreopithecus*. This ape lived in the Late Miocene (7–9 million years ago) and was endemic to a Mediterranean island whose fabric now forms parts of the Italian region of Tuscany. *Oreopithecus* was, in its own way, a biped, so much so that its hands were sufficiently free to allow for a precision grip, in which the tips of the fingers and thumb can be pressed together, allowing fine manipulation—something often assumed to be exclusive to toolmaking humans.[32] But *Oreopithecus* was a very distant cousin of hominins, not an ancestor.[33] Its bipedality was not a harbinger of technology, holding babies close to its chest, or anything else. The free hands of *Oreopithecus* were not, as far as we know, employed in making tools, thereby refining "planning depth," swiftly followed by the conquest of the earth. Whether bipedality in the species was connected with sexual display will probably remain forever unknown. We do know that bipedality did not save it. As far as we know, *Oreopithecus* remained confined to its island home, where it quietly became extinct. For *Oreopithecus*, bipedality was a trait as individual as any other variety of ape locomotion, not the first step in some progressive path of transformation between Ape and Angel.

And the same is true for us.

8: *The Dog and the Atlatl*

One of my favorite items of technology has an ancient pedigree. It's a springy, flexible rod about fifty centimeters long, with a handle at one end and a cup at the other to hold a tennis ball. I use this to throw, with ease, tennis balls for my dogs to chase and fetch, much farther than I could throw them unaided, even with great effort. If on any given day you can't find me at home or at the office, try the beach: there you'll find me using the ball thrower to throw balls for my dogs to retrieve.

The principle is simple—by lengthening my arm, it increases leverage. By expending the same force, the ball leaves the end of the atlatl with greater velocity than it would from my unaugmented arm. Devices like this have been used to throw darts and spears for tens of thousands of years. Mine gets a modern makeover in that it's made of plastic rather than the wood, bone, or antler of the originals. The material aside, the atlatl or "spear thrower" is one of human technology's more enduring design classics.

Technology needs a definition. To most people, I suspect, the word "technology" conjures images of complicated machinery or modern electronic hardware. But such modern technology, even if apparently changed beyond recognition, is really a compressed combination of many much simpler technologies.

Take, for example, the iPad on which I wrote much of the draft for this book. It is made of plastic and metal. The plastic comes from the chemical processing of petroleum, reliant, at root, on nineteenth-century chemistry based on eighteenth-century engineering. The metals are occasionally exotic, but the basics of mining and metalwork go back to antiquity. The electronics in my iPad are based on VLSI chips (very large scale integrated) of what twenty years ago were called microprocessors, themselves condensed versions of transistors (invented in the 1950s), and before that, vacuum tubes that go back to a nineteenth-century fusion of once-separate technologies such as metalwork and glassblowing. The programming that allows me to write on the iPad has a distant ancestor in punched-paper tape used to program vast computers made of arrays of vacuum tubes: and at the business end I use writing, a system of symbols for the recording of ideas whose concept is (by definition) as old as history. When looked at critically, even the most advanced technologies used today by human beings are variations and combinations of earlier, simpler ones.[1] In any case, the kind of technology in use for most of human existence has been of the order of the atlatl: simple machines that allow one person to exert greater force than he might have achieved unaided. Give me an atlatl long enough, as Archimedes might have said, and I could throw a tennis ball from here to next Tuesday.

How does one define technology? One definition might indeed encompass all those many objects that people create to do things they might not be able to achieve (or might achieve less well) unaided—things like the atlatl. Such a definition extends well beyond objects we recognize as tools or weapons, to the clothes and dwellings that allow us to live in places that humans might not otherwise have penetrated, to the pottery used since antiquity to transport or contain such items as nuts and grain, fire and water, and in which food might be cooked.

Cookery in particular is believed to have had a profound influence on human anatomy, physiology, and social structure.[2] Cooking food breaks down hard or woody tissues, neutralizes toxins, kills potentially harmful bacteria and parasites, and makes more nutrients available to the diner. This increase in efficiency meant less time spent foraging and digesting, allowing more time for social interaction—aside from the fact that cookery tends to be a social and sociable activity in itself.[3] Some scientists think that cookery was followed by a reduction in the mass of the jaws, teeth, muscles, and digestive tract, and perhaps an increase in the mass and capabilities of the brain. It is not trite to suggest that humans have been modified by their own technology.[4]

This definition of technology—those things we create outside our own bodies that allow us to do things we could not have done unaided—also encompasses things that we might not describe as technological at all. Technology might be said to include such imponderables as legal codes and financial structures. Laws make it easier for people to live together harmoniously. Financial structures—from coins and notes to credit derivative swaps and futures contracts—allow us to exchange goods without having to physically carry them around ourselves. Laws and money are not technology in the sense of tools you hold in your hand. Rather, they represent social conventions. The twenty-pound note in my pocket doesn't actually *do* anything—it is merely a promise by the Bank of England to underwrite any transaction made with it to that value. Once upon a time such transactions were backed with a real commodity (gold), but no longer. My twenty-pound note is itself a real *thing* created by the technology of printing, but represents another sort of technology based on social contract. Technology, therefore, includes things that we might otherwise regard as social conventions rather than physical objects.[5] "Money" only has "value" inasmuch as we all agree that it does. Therefore, the importance of such things as money—and with that, wills, contracts, and treaties—lies not in their physical form but in the ideas they represent. If this seems somewhat rarefied to be technology, consider that people are given life or condemned to death as surely by abstractions such as treaties and the passage of money as they are by more concrete examples of technology such as antibiotics or nuclear weapons.

If such intangibles as money can be regarded as technology—spoken of in the same breath as, say, swords and plowshares—then perhaps the earliest and most enduring example of technology is something one might not regard as technological at all: the domestic dog.

Not for nothing does the dog bear the soubriquet of Man's Best Friend. For tens of thousands of years, dogs have made human beings safer, helped them herd other domestic animals and hunt wild ones.[6] In the old days, when, to paraphrase Jared Diamond,[7] we didn't do all our foraging in supermarkets, I'd have gone hunting with my bone or antler atlatl and some spears, and a fast-running dog to chase down the prey and retrieve the kills. These days, my dogs and I reprise the same activity, purely for pleasure and exercise, with a plastic atlatl and some tennis balls.

Modern dogs have been bred for a variety of purposes, some quite remote from what we assume to have been their original uses, such as hunting or retrieving game (Heidi, my golden retriever), ridding campsites of rodents (Saffron, my Jack Russell terrier), guarding against intruders (both), and herding livestock (neither). Modern dogs are used in such sophisticated tasks as helping blind and deaf people get around busy cities, rescuing people washed into the sea or buried under rubble, sniffing out explosives and contraband, but perhaps most of all to provide companionship for people.

My friend Brian Clegg, a full-time author, says that the most important piece of equipment a writer can own is not a computer, nor even a pen, but a dog. Writing is a solitary business, and a dog provides company without being intrusive. One can always postpone a trip to the gym, but those appealing doggy eyes and waggy tail have a way of persuading the writer to take regular breaks for exercise, whatever the weather, during which the writer can think about what he has just written and plan the next bit.

How can a dog count as technology? It fulfills my definition in that it allows humans to do things that they might not have been able to have achieved unaided (hunting, herding, sniffing out drugs, eyesight to the blind, and so on). My definition also specifies that technology is created for its purpose. The domestic dog is definitely a creation that would not have evolved naturally, having been modified quite extensively in both behavior and appearance from its ancestor, the wolf. Most cases of domestication involve humans breeding and selecting animals and plants in true Darwinian style to optimize them as producers of food. Dogs, however, are a special case of one social carnivore being domesticated by another social carnivore for mutual benefit—a kind of symbiosis.

How far back does human technology go? We know for certain that technology antedates modern *Homo sapiens*. The earliest tools that can be recognized as such are chipped pebbles from Ethiopia that date back 2.5 million years, although it is possible that hominins were using sharp stones to butcher carcasses of other animals (such as antelopes) as long ago as 3.4 million years ago.[8]

The earliest stone tools don't look like much, and it takes an expert eye to tell them apart from pebbles broken by natural causes or by accident. However, some stone tools, notably the hand ax—that canonical item in the Stone Age toolbox—are objects of remarkable beauty, and show evidence of extraordinary craftsmanship to rival anything one might find in, say, the workshop of a trained cabinetmaker or stonemason today.

But human beings are not unique in their manufacture of objects that are both useful and beautiful. In fact, living things have been creating such objects for almost as long as life itself has existed. Stromatolites—cushion-shaped domes that survive in salty lagoons—are the sculpture-like objects created by simple, single-celled cyanobacteria (blue-green algae). Around 3 billion years ago, when such creatures were the most complex forms of life on the planet, stromatolites were common and formed the first reefs. Today, reef-forming creatures such as corals (simple animals, related to sea anemones) create complex and beautiful structures, external to their own bodies.

On land, social insects create complex and well-engineered structures in which they live—one thinks of honeycombs with their regular, hexagonal cells made of wax. The towering nests of termites dwarf their creators in the same way that skyscrapers dwarf their human architects—and have air-conditioning systems to rival any created by human designers. Birds, too, make nests, some highly elaborate: think of the nests of weaverbirds, hanging from the branches of African trees, or the elaborate stages created by male bowerbirds on which they display to prospective mates. At least one bird species—the New Caledonian crow—makes and uses tools that conform to my definition of technology, modified leaves for use as probes.[9] Anthropologists are even beginning to recognize that apes, chimpanzees in particular, have rudimentary technologies, in which they use stones to crack nuts, or strip leaves from stalks to allow them to probe anthills. Some of these technologies even have a "cultural" dimension.[10] That is, they vary from one group of primates to another according to the learned traditions of each.

Such discoveries have led to the emerging discipline of "primate archaeology"—the excavation of occupation sites used by primates other than humans, in order to learn about the history of their technological and cultural traditions.[11] This discipline is beginning to offer a much-needed comparative perspective on the relationship between technology and human history, by first admitting that technology is not unique to humans. After all, the first hominins to make tools were not humans in any sense we'd allow today. The first tools used by hominins would have been neither more nor less sophisticated than those used by modern chimpanzees. All this leads to a perhaps obvious question—why should beautiful, useful technology created by humans imply a maker any more intelligent or deliberate than (say) a crow or a weaverbird? A termite or a stromatolite? All such creatures fulfill my definition of technology in that they create things for use outside their own bodies that allow them to do things that they might not have done unaided, irrespective of their cognitive abilities.

To reserve technology for humans requires the inclusion of such imponderables as "planning depth," assumed to be an exclusively human attribute, which can be defined as the ability to plan for future eventualities. For example, before I can paint this wall, I shall need to open this tin of paint, but before I can do *that*, I shall need to go to my shed and find a screwdriver to use as a lever for prizing open the lid.[12] Should "planning depth" be included in any definition of technology? I suggest not, because the concept creates more problems than it solves. We know, for example, from behavior experiments, that some animals that use technology—notably crows—do indeed plan ahead in a way that is indistinguishable from human behavior in similar circumstances.[13] In which case the attribute of planning depth is not unique to humans.

For other animals, the creation of structures outside the body is presumably instinctive, that is, hardwired, rather than learned by observation—or even a completely incidental by-product of metabolism. This is probably true for all those organisms such as corals or stromatolites, which lack what we would recognize as a brain. However, this might be a distinction without a difference, because of the a priori assumption, not stated in my definition above, that brains are an absolute prerequisite for technology. But if a creature, whether or not it has a brain, modifies its environment to allow it to live where it otherwise might not, can that not be regarded as technology? Does the hidden assumption that brains are necessary prejudice one toward the view that technology is something we humans award ourselves, exclusively, because of our privileged view, and our tendency to think that we are the culmination of all organic achievement? Were the myriad polyps responsible for the Great Barrier Reef to be polled on the issue, they might with justification say that theirs is a more magnificent achievement than anything made by Man. Polyps, though, have no brains—but does it matter? Is it not easier to judge what is and what is not technology by easily observable results rather than on the motivational states of the makers, which are much harder to fathom? Because it is easier to infer motivation in fellow humans than in nonhumans, our view on technology will quite naturally be prejudiced to it being a humans-only activity—but only so long as concepts such as "planning depth" are considered as definitive.

Brains are, however, definitely required for "planning depth,"[14] whether or not this planning is applied to the manufacture of objects outside the body. However, such a concept runs into a mire of philosophical problems as eloquently discussed by Daniel Dennett in his book *Consciousness Explained*. Briefly, the idea that we imagine the world outside our heads as a dramatic performance that we are somehow observing through the windows of our senses—what philosophers call the "Cartesian theater"—is an illusion easily bruised by any number of experiments. So, whereas I might imagine a little picture of myself going to my shed to get the screwdriver I need to lever open the lid of that tin of paint I mentioned earlier, such a drama is a rationalization after the fact of a host of disparate thoughts and impulses. If there is no such thing as the Cartesian theater, there can be no such thing as planning depth. If there is no planning depth, there is no necessity to infer that brains are required for technology at all—in which case one is left with the definition of technology with which I started, which therefore must include, along with my iPad and the probes created from leaves by those clever crows from New Caledonia, the insensate creations of bees, corals, termites, and even cyanobacteria, without reference to the internal motivational states (if any) of the creators.

This problem of planning, or intention, crops up whenever we think of the hand ax, that quintessential example of Stone Age artistry. Although it's always hard to be sure, hand axes are generally associated with a particular species of hominin, *Homo erectus*, and are examples of a toolmaking tradition or style known as "Acheulian" or "Acheulean," after the site in France called Saint-Acheul, whence such tools were first described as such. The earliest known tools in the Acheulian style come from Kenya and are almost 1.8 million years old.[15] As *Homo erectus* spread around the Old World, hand axes went too. They have been found from Spain to China, England to Indonesia. Although some later hominins such as *Homo heidelbergensis* and Neanderthals adopted and modified the hand ax, the basic plan remained more or less the same for 1.5 million years,[16] variations dictated more by the different materials used in hand-ax manufacture rather than any change in tradition or culture.

The essentially unchanging nature of hand axes suggests that the techniques used to make them were not entirely learned or taught, but were to some extent hardwired. Even if *Homo erectus* adults taught them to their offspring, there was absolutely no conception that hand axes could be made in anything other than the prescribed way. And if, in the sequence of steps used to make a hand ax, a blow went awry, spoiling the blank, the maker wouldn't shrug his shoulders and make the best of a bad job, converting what might have been a hand ax into a scraper, a Large Hadron Collider, or a hi-fi cabinet. No, he would start all over again with a new blank. There are some Paleolithic sites at which hand axes have been recovered in great abundance, in various stages of manufacture. These sites are exceptional—but the exceptions still require some kind of explanation.

Perhaps most tellingly, there is much argument about what hand axes were for.[17] Although they are very beautiful,[18] they are in many ways impractical. A knife or chopper made of stone is easier to hold in the hand if some part of it retains the original, smooth stone surface—but hand axes are flaked all the way round. The raw edges of a cut flint are extremely sharp to begin with, but the edge soon dulls. So if you are going to make a chopper to smash bones or a knife to slice through flesh, it's quicker and easier just to strike a flake and get on with it, rather than commit to the immense artistry required to make a hand ax. So what else might hand axes be for? Currency? Symbols of status? Sexual display?[19] It's impossible to know. What we can say from the evidence is that hand axes represent the kind of stereotypical behavior associated with other examples of animal technology, such as the nests made by birds, woven with great skill but always in the same general way. Whereas it is true that some aspects of behavior that seem stereotypical are to an extent learned—birdsong is a good example—the songs of birds are always pretty much the same and characteristic of each species. The might have been true for hand axes and *Homo erectus*.

And yet *Homo erectus* looked very much more like us than any kind of bird. Is it fair to dismiss his works as the products of—for want of a better word—instinct? After all, *Homo erectus* is thought to have tamed and used fire.[20] The discovery of stone tools at least a million years old on the island of Flores[21] shows that *Homo erectus* was capable of crossing stretches of deep ocean out of sight of land—something that might well have involved a great deal of organization and planning. Yet we know that many animals less obviously endowed with intellect can cross stretches of open ocean by accident. In *The Wisdom of Bones*, a detailed look at the life and times of *Homo erectus*, Pat Shipman and Alan Walker conclude that *Homo erectus* would have had no more spark of what we might call "humanity" than any canny social savanna predator, such as a lion or a hyena. Studies on the development of *Homo erectus* teeth and skulls show that these creatures grew rapidly from infancy to adulthood, rather in the manner of apes, and lacked the extended period of growth called "childhood" during which a young modern human learns social skills from adults.[22] To be sure, hyenas and lions teach their cubs about the ways of the world, and we might expect *Homo erectus* adults to have done the same. But that does not mean that the knowledge they imparted was any less hardwired, nor that the process of teaching and learning is not in itself stereotypical behavior.

The million-year stasis of hand axes stands in stark contrast with the technology associated with *Homo sapiens*, especially after about 45,000 years ago when the first modern humans appeared in Europe.[23] If the technology of *Homo sapiens* can be summed up in one word, it is "change." Modern human technology is always changing and developing as humans learn from their mistakes, never discarding errors but learning from them to improve the old or invent something entirely new. In the light of modern human technology, the technology of *Homo erectus* is not technology as we understand it today—at least not conventionally.

The shock one experiences when looking at a Stone Age cave painting or Venus figurine is that of recognition—that after millions of years of chipped pebbles and hand axes of unknown purpose, we can recognize the product of a mind that is distinctively human.

And that's a worry, because it introduces that inescapable referential bias that plagues any study to do with human evolution—that we are both the subject and the object of study, and will naturally know (or think we know) more about ourselves, and how our minds work, than of the minds of other creatures, including extinct hominins such as *Homo erectus*. It is only us, looking backward from our perceived high estate, that look at stone artifacts and immediately assume that they must have stemmed from the same creative, artistic, practical urges that we experience ourselves. That the motivations of *Homo erectus* might have been alien to our way of thinking seems an affront until one looks at the evidence dispassionately.

Does this contrast—between the works of modern humans and those of *Homo erectus*—elevate human technology to some kind of special status? The constant change and invention that is typical of modern human technology seems to mark it out as something quite different from (say) a coral reef or a termite mound. Well, yes—and no.

The change and invention is linked with the long childhoods of modern humans, but it remains the case that human children are taught, in much the same way as the young of other animals are taught, in a stereotypical way that is determined in part by the physical constraints of brain growth and development. For example, modern humans have an innate capacity for language—any language—without being taught in any conscious way, but must also be exposed to it at a certain time in infancy for it to develop properly. And the fact that we can't grasp the purpose of the *Homo erectus* hand ax shows that despite its protean character, human technology is not infinitely malleable. It is fundamentally limited by our capacity for understanding or conceiving what is possible, according to our senses and how our brain interprets what they are telling us. We, like *Homo erectus,* have our boundaries. To us, they seem infinitely far away, over the intellectual horizon. But who's to say that *Homo erectus* wasn't similarly overoptimistic about his own limitations?

It is entirely natural for us to think about our capacity to make beautiful things and on that basis ascribe to ourselves capacity for forethought that we deny every other living thing. I hope I have shown that this self-justification neither fits with a proper reading of evolution, nor is it fulfilled by the evidence. Unfortunately, our understanding of human evolution has become forever muddied by such self-aggrandizement. When Louis Leakey discovered the remains of a distinctive fossil human, which he called *Homo habilis,* "handy man," the name was a direct reference to the discovery, in the same strata as the fossils, of very primitive stone tools. That *Homo habilis* had made the tools was not to be doubted, or so Leakey thought. The fact that remains of another extinct human, *Zinjanthropus* (now *Paranthropus*) *boisei,* were also found in these strata, was played down, because *Paranthropus* has a smaller brain than *Homo habilis.* The larger brain was meant to go with the tools. The tools must have been made by a large-brained creature, whose mind was stuffed full of what came to be called "planning depth."

This circular argument has been the source of no end of trouble, not least that the species itself was defined, in part, by a technology it was supposed to have created, when there was no certain way of linking tools and toolmakers. Ever since the 1960s, and with the discovery of more fossils of *Homo habilis,* people have worried about how to recognize fossils of *Homo habilis* should they find them,[24] given that a defining feature of the species is a kind of behavior that not only does not fossilize, but which might not be unique to humans or the genus *Homo*. Some have even wondered whether *Homo habilis* should really be regarded as another form of *Australopithecus*.[25] In recent decades, species such as *Australopithecus garhi* and *Australopithecus sediba* have been described as having some claim to close relationship with *Homo*.[26] *Australopithecus garhi*, from Ethiopia, has at least as good a claim on the authorship of the earliest known stone tools as any member of *Homo*.

It was the brain argument that was the real issue. Anthropologists looked at the skull of *Homo habilis*, painted it against the canonical picture of a progressive increase in brain size, and decided—retrospectively—that there was a brain size above which some kind of mental light would switch on, and the ape would become, if not an angel, then an artisan. The problem is that there is no simple connection between brain size and intelligence, a topic I'll explore in the next chapter.

9: *A Cleverness of Crows*

If, after all that, I haven't convinced you that there is nothing special about human beings that merits some elevated position on the top of nature's tree, I know someone who might. That person is Nicky Clayton, professor of experimental psychology at the University of Cambridge and fellow of the Royal Society of London. She is the only Cambridge professor I know who arrives at work in a bright red dress and high heels. Scientist by day, dancer by night, she is an expert at the Argentine tango.[1] And she spends a lot of time with birds of the family Corvidae—crows and jackdaws, jays and ravens. A single visit to Professor Clayton's aviary should convince you that intelligence—if it stands for anything at all—is not confined to human beings.

Clayton and her colleagues are learning to understand what goes on in the minds of nonhuman species. Corvids are excellent subjects. They are small, proverbially crafty, easy to keep in captivity, willing participants in experiments, often highly social, and there are lots of different kinds. This last means that results can be compared between species with different types of social behavior but equivalent apparent intelligence and brain size. This is something that can't be done with humans, as we have no extant relatives that resemble us in intellectual facility or brain size. Whatever one means by "intelligence," the great apes seem to have much less of it than humans. But they also differ markedly in social behavior from humans (and one another), as well as in brain size, which could both be factors. If apes were more sociable, or had bigger brains, would they be as "intelligent" as humans? Studying the variety of crow species—from ravens to jackdaws to jays to plain old crows—has the potential to adjust for the interaction (if any) between social behavior, brain size, and intelligence (and I'll be returning to that subject, too). I suspect that we'd have a much more nuanced view of our own importance were Neanderthals or Denisovans still around with whom we could compare notes.

In a long series of experiments, Clayton and her colleagues, as well as researchers elsewhere in the world, have shown how various species of crow are capable of many feats of intellect usually associated only with human beings. As I noted in the last chapter, the New Caledonian crow snips and shapes leaves to make tools every bit as useful as the probes chimpanzees use to extract termites from nests—or early hominins made for butchering meat. More remarkably, crows can use tools to make other tools to achieve a task.

The cleverness of crows is proverbial. Everyone must have seen, by now, videos showing how crows leave nuts in roads, waiting for them to be cracked by the wheels of passing traffic—and the trick of those especially clever crows that leave nuts on pedestrian crossings, allowing the crows to retrieve the spoils without getting run over.[2] In her lab, Clayton showed me a video showing how, when a crow is confronted with a morsel floating in a beaker of water but too deep for it to reach, the bird will use stones nearby to displace the water, raising the morsel to the surface and allowing it to be reached. To do this, the crow had to be able to appreciate the various properties of materials, such as that the food scrap floated, even when stones were thrown in the water; that stones would fall to the bottom; that stones displaced the water (equivalent to Archimedes' "Eureka" moment); that the water would rise up the beaker, carrying the morsel of food. Not only that, the bird would have had some concept of itself throwing the stones into the water to achieve the desired outcome. So, not only can crows think things through, they are capable of thinking through what they are thinking through. And they are also capable of thinking through what *other* crows are thinking through.[3]

To me, the most remarkable fact about crows is that theirs is a kind of intelligence that we can recognize—the calculation and the craftiness are things we see in ourselves. I do not think one is going too far by saying that the minds of crows work in a similar way to ours. In many ways, the human mind has more in common with the minds of crows than with our closest cousins, the apes.

If this is true, it is remarkable, because crows and humans have brains that evolved entirely separately, along completely distinct pathways.[4] The common ancestor of crows and humans was some kind of reptile that lived more than 250 million years ago, and would not have had enough brains to write home about. As a result, the human brain—and that of other mammals such as primates, dogs, whales, horses, and so on—is made rather differently from that of crows.

This is an important insight in the context of this book because, once grasped, it shoots a huge hole in the idea that what we think of as the human mind must necessarily have evolved from earlier hominins simply by virtue of the fact that they were hominins, and had an evolutionary heritage that would have demanded progressive cognitive improvement in that lineage alone. It forces us to look at what we and crows have in common, to the exclusion of apes—and, from that, helps us understand the evolution of intelligence in general terms, not just in our own evolutionary lineage. All such similarities must very greatly be concerned with behavior rather than anatomy, as human brains and crow brains are wired differently, and crows don't have the hand-eye coordination sometimes thought of as having been instrumental in the evolution of the human mind.

What humans and crows (and many other birds) have in common is an active social life.[5] Unlike apes, which are solitary or live in small groups, humans and birds tend to live together in large groups in which relatives of various ages mix together with less familiar individuals. They tend to learn from one another, but they are also competitive. They have a level of technological sophistication that outranks, in concept at least, anything seen in apes (even allowing for the fact that crows don't have hands). Human and bird societies are cohesive and complex, and prone to a certain amount of internal discord and deceit. As I discussed earlier, cuckoldry is common in birds that are apparently monogamous, as it is in human societies, and this circumstance might, paradoxically, keep societies together, as birds will seek to keep an eye on not only the fledglings in their own nests, but those in the nests of their neighbors.

There's no doubt that the minds of crows are comparable in capability with those of humans, and have much the same flavor, for all that crows have no language, no hands, and brains the size of berries. A short visit to Clayton's lab should dispel any notion that intelligence is necessarily all about brain size or hand-eye coordination. That we can recognize the same phenomena in creatures as distantly related to us as crows suggests that what we think of as intelligence might have less to do with the physical structure of brains in isolation, than with the complexities of social relationships quite irrespective of form. If we find intelligent aliens, we'll recognize them, too. They'll behave just like we do.

Intelligence, however, does seem to have something to do with the mass of the brain relative to that of the rest of the body, irrespective of the brain's actual size. This measure is called the encephalization quotient, or EQ.[6] Animals with a high EQ have large brains relative to the size of their bodies. Crows have small brains, but they also have small bodies, so their brains tend to be relatively large compared with those of less clever birds of similar mass. That is, crows have a higher EQ than, say, pigeons or chickens. It is also true that human beings have a much higher EQ than mammals of comparable mass, and considerably higher than those of apes. Even bearing in mind the dangers of coming to a narrative, progressivist conclusion, the human EQ has increased rapidly and markedly over evolutionary time. Compared with those of apes, it is off the scale: the relative and absolute increase in brain size has been greater than for any other organ or organ system.[7]

Not only is the modern human brain large, it consumes a disproportionate amount of energy. Even though it is large in proportion to our mass when compared with brain masses in other animals, it still constitutes only between a fiftieth and a hundredth of the mass of the body—yet it consumes one-sixth the energy. The brain's expansion has distorted the skull so grotesquely that even though human babies are born in a relatively immature state, the hugeness of the infant's head puts a mother's life at risk. No doubt about it, the human brain is big. Bothersomely big. So it must be doing something. But what? What is the human brain for?

By now you should be able to recognize that proposals of purpose should be treated with caution. Just because human brains are big does not in itself necessitate a simple explanation for such disproportionate size. Human brains might have evolved for better hand-eye coordination, for example—but that's one of those circular explanations that I hope we've put behind us. In any case, work on crows disposes of that idea quite nicely—although quite a large amount of brain is devoted to coordinating the fine degree of dexterity of which human hands are capable, crows have large EQs and can make tools, and they must make do with their beaks.

It is more likely that the human brain evolved to be as large as it is by virtue of a number of different circumstances that interacted with one another—sometimes reinforcing, sometimes opposing—over the course of human evolution. The evolution of the human brain, like the evolution of anything else, must be thought about in terms of Darwin's tangled bank, rather than the misreading of evolution as linear, progressive, and governed by purpose.

But first, to the brain's bothersome bigness. A clue to why the human brain is so big can be found in the timing of human brain expansion in evolution.

It took off sometime after the evolution of *Homo erectus* but before the appearance of Neanderthals.[8] Significantly, this expansion occurred long after the invention of tools and technology.[9] When Louis Leakey and colleagues announced the discovery of *Homo habilis* in 1964, the whole idea of technology was linked with intelligence and brain size, leading to a long and fruitless discussion about the size of brain a fossil hominin ought to have before it could be considered either intelligent or technologically capable—as if, when the brain exceeded a certain size, a mental light would switch on and, like the apes in *2001: A Space Odyssey* confronted by a monolith, they would be catapulted into a new realm of cognition.

That period of prehistory—between 1.5 and 0.5 million years ago—is particularly murky. The world was inhabited by one or more species of hominin, collectively referred to as *Homo heidelbergensis*,[10] which presumably evolved into Neanderthals in Eurasia, *Homo sapiens* in Africa, and possibly other species in China and elsewhere. We know that *Homo heidelbergensis* individuals were big and beefy, which alone would have contributed to their large brain size. But we also know that they were technologically fairly accomplished. Well-fashioned wooden spears dating back some 400,000 years ago, and miraculously preserved in peat in Schöningen, Germany, were arguably made by *Homo heidelbergensis*.[11] They look like well-balanced hunting javelins, but would have required a person of some stature to use them effectively. Another habit possibly started by *Homo erectus* and continued by *Homo heidelbergensis* was the controlled and deliberate use of fire, which would have led, very quickly, to the invention of the barbecue. As I discussed in the previous chapter, some have suggested that the invention of cuisine contributed in no small measure to the evolution of the large brain in humans.[12]

Whereas herbivores have relatively large guts and rather small brains, the opposite seems to be true in carnivores. The reason seems clear. Plants don't take great skill to eat, if you are doing nothing more complicated than grazing or browsing, but they require formidable powers of digestion. Plant cells hoard their nutritious innards behind tough walls of cellulose, digestible only by bacteria living in the gut. Plants also contain many substances that are poisonous, and these need to be neutralized (plant poisons, a kind of defense against herbivory, form the basis of many drugs used today). Even then, the amount of nourishment one gains from a given mass of plant is rather small. Hence the large gut, and the enormous amounts of time that herbivores spend eating. Carnivory, on the other hand, requires a certain skill, as one's next meal might have a clue that you are chasing it, and will therefore be prepared to take evasive action. This alone suggests that carnivores require proportionately larger brains than herbivores. In addition, meat is much more easily digested than vegetation, meaning that much less time needs to be spent eating—and much less investment is required in digestive machinery.

It is possible that hominins started to incorporate meat as a large proportion of diet with the evolution of *Homo erectus*. This is evident from gross anatomy. Australopiths had cone-shaped, flared rib cages suggestive of a large, pot-bellied gut. *Homo erectus* and later hominins had a cylindrical rib cage, suggesting a trimmer physique and a smaller gut. The evolution of carnivory allowed hominins to invest more in energetically expensive brain tissue at the expense of the gut.

This "expensive tissue hypothesis" makes intuitive sense.[13] It has, however, been dented by recent work showing no strong correlation between massive brains and lightweight guts in a large variety of mammals.[14] What the data do show, however, is a link between brain mass and mass of body fat. Animals with small brains tend to have more fat reserves. Animals with larger brains tend to make do with less fat. The rationale is that having a store of fat is a hedge against thin times ahead, especially if you don't have the smarts to go out and find more food. Animals with smarts, however, can presumably think about how and when to get their next meal, and therefore gamble on carrying less fat.

The exception to the rule, as it happens, is humans, which have both large brains and large stores of fat. This could offer a belt-and-braces solution to the prospect of starvation, but there are other possibilities. Advocates of the "aquatic ape" hypothesis discussed above might note that whereas humans seem anomalous among land animals, having large brains and large fat stores together is the rule for marine mammals such as seals and whales. The "aquatic ape" hypothesis, however, doesn't exclude the very gender-specific patterns of fat storage seen in humans, which might indicate that it has more to do with sexual selection than natural selection.

The tendency to carnivory might have been enhanced by cooking, which does a number of remarkable things to food and might have exerted some influence on human evolution. So, what does cookery do, apart from offering a stream of entertaining TV programs from celebrity chefs?

The first and perhaps most important thing is that by breaking down tough proteins and fibrous materials in food, cooking releases more nutrients per unit of mass than one can extract by eating a morsel raw. In short, cooking is a kind of predigestion. If we cook food before we eat it, we need devote less time and energy chewing and digesting it once it has passed our lips. This has a number of important implications. By using a given amount of resource more efficiently, cooking allows people to be bigger—that is, more massive—than they would be otherwise. Some increase in brain mass would be a simple side effect of an increase in body mass, irrespective of any other cost or benefit. Cooking also neutralizes any toxins and kills parasites in food, so that people eating cooked food might live longer and be healthier than those subsisting entirely on raw food.[15]

There would be changes in shape as well as size. Modern humans have smaller teeth than many earlier hominins, and this has also been related to cooking. If food has been softened by cooking, one needn't expend any more resources than necessary building a big and complicated digestive system. This applies to teeth and jaws as well as guts, and to the muscles that open and shut our jaws. Much of the force of the bite inflicted by a dog, say—or a chimp—comes from the chewing muscles on each side of the head. These muscles run from the sides of the jawbone, thread beneath the zygomatic arch (cheekbone), and fan out on the sides of the head, anchoring at a crest at the very top. Animals with big chewing muscles—such as dogs and gorillas—often have such a crest running along the midline of the skull. Hominins such as *Paranthropus* had teeth like tombstones, powered by big, thick chewing muscles—we can tell this from the prominent head crests and the widely flared cheekbones that made way for the muscle mass. Modern humans, though, have none of these things, and the relatively weak chewing muscles get only as far as the sides of the head before petering out. The skull roof is smooth and not clothed in muscle. This is one reason why modern human skulls are globular, with no crests or other prominent signs of muscle attachment.

The smallness of human chewing muscles has been linked with a particular genetic mutation found in humans but not other mammals.[16] Could this mutation have played a part in the expansion of the human skull and the weakening of these muscles? Could natural selection against this mutation have been weakened in a population of novice chefs and aspiring gourmets, allowing it to spread? Irrespective of the mechanism, the reduction in muscle mass, and the exposure of the skull, would have removed a possible external constraint on skull growth. The growth of the skull in babies and small children is intimately linked with brain growth.[17] The brain drives the expansion of the skull roof, from which it seems possible that the reduction of jaws, teeth, and their associated musculature might have been connected with the further expansion of the human brain, beyond any expansion that would have accrued as a simple increase in body mass as a consequence of the nutritional benefits of cooking.

The expansion of the human brain starts early, well before birth, so that the size of the fetal brain is so great that delivering babies is a serious health risk for human mothers. The brain size of human newborns seems to be at the top of the permissible range. We can tell this because the brain continues to grow and develop in the human infant for far longer after its birth than in apes. As a consequence, human babies are born in a relatively helpless, premature state compared with the babies of apes and many other mammals, and take a very long time to mature. As every parent knows, bringing up children requires a vast amount of effort and resources, and is far less onerous if one can share the burden. It is known—or at least strongly suspected—that women who can call on relatives to help will raise children more successfully than those for whom help is unavailable. In traditional societies, at least, a great deal of help comes from a mother's older female relatives, especially her own mother. This so-called grandmother hypothesis might explain another otherwise puzzling feature of human biology—the menopause.[18]

In virtually all creatures, the evolutionary imperative means that having been born, one should grow up as quickly as possible, reproduce early and often, and then die. Very few creatures live for very long past reproductive age, but humans are a marked exception. In human females the process of menopause marks a definite shutdown in reproductive capacity—after which the individual can expect to live for several more decades.[19] An explanation, which seems to have some traction, is that by ceasing to reproduce herself, a female is then better able to assist the reproduction of her daughters.[20] Because human babies take so long to mature, one can imagine selective value in both prolonging human female life span and introducing a definite cessation of reproductive effort in order to care for one's grandchildren, as well as one's own children—rather than producing more children who will compete for resources with one's grandchildren. And, given the high risk of death in childbirth as a function of the large brains of babies, the odds might be stacked in favor of ceasing reproduction to care for grandchildren rather than incurring the risks entailed in producing more young of one's own—risks that might become greater with age.

The menopause has other effects, too. It automatically provides a stratum of society that other animals simply do not have, which is a cadre or caste of older individuals whose existence is not predicated solely on their own reproduction. By offering stores of knowledge that can be passed on to younger members of a group, elders make societies more cohesive and offer the potential for such groups to become more complex.

Could all this—the expansion of the human brain, the helplessness of human infants, and the growth of society—all be related to the invention of cooking? Even were one to regard the above with a laudable skepticism, the fact remains that cooking is in itself a social and sociable activity. People gather round the hearth, and there can be few social, ritual, or religious occasions that do not revolve around the provision of food and drink, sometimes after specified periods of abstinence and fasting. As religious offerings were once edible (and suitably cooked to send the savor of cooked food to heaven), no birthday party is complete without a cake, just as Christmas isn't Christmas without a Christmas pudding, Thanksgiving without its turkey, nor Easter without its eggs.

And where people come together to eat and drink, they gather to do deals, choose mates, play music, sing, dance, and swap stories. In humans, at least, cooking facilitated our need to be social. Therein might lie another clue about brain expansion. All the animals we know that have large brains, large EQ, and behavior that we humans recognize as "intelligent" are also social. Brain size, intelligence, and social life go together.

Or do they?

Before leaving this chapter, I'd like to take a quick look at Neanderthals. These were hominins with brains larger, on average, than those of *Homo sapiens*, and of a comparable EQ. They were capable chefs, made tools, and perhaps even had an inner spiritual life. Neanderthals lived on Earth for around 300,000 years but left without making a mark. *Homo sapiens* has been around, so far, for around 200,000 years, and in just the past 10,000 years has come to dominate the planet's ecology and resources. One is entitled to ask why.

Clearly, brain size isn't everything. But what several recent lines of research have indicated is that Neanderthals were a lot less sociable than modern humans. They tended to live in smaller groups, had smaller home ranges, and were therefore less likely to come across other members of their own species.[21] Neanderthals were, in this respect, rather like modern great apes. Perhaps the earth was inherited not by the creatures with the biggest brains, or the most intelligence, but with the busiest social calendars.[22] The invention of agriculture between around 10 and 20,000 years ago created a situation in which humans were forced to live in fairly large, concentrated groups. The first villages were less like enlarged nests of gorillas or the expanded ranges of chimps, and more like rookeries.

I've used the word "intelligence" several times, but without attempting to define it. I have, however, compared human intelligence with crow intelligence, suggesting that crow intelligence is something we can intuitively recognize. But is it possible to define intelligence on its own terms, without reference to the animals in which it is found? Can it ever be isolated from cultural or developmental context? The application of such things as IQ (intelligence quotient) testing is too parochial for the remit of this book, which aims to take a wider, more zoological view. When applied to the wider world of life, one might propose that intelligence is a rough measure of the speed and efficiency of information retrieval, perhaps combined with a way of generalizing this information so that it can be applied in novel situations—something like Spearman's original concept of "general intelligence,"[23] derived from the observation by the pioneering statistician Charles Spearman, a century ago, that schoolchildren who were good at one subject were likely to be good in others, too, reflecting an underlying intelligence unrelated to subject-specific capabilities.

Importantly, this rough-and-ready definition of intelligence says nothing about brains, how big they are, what they are made of, or how they are wired up. This is important, for it allows us to look at creatures whose brains are made very differently from ours, and estimate what intelligent creatures have in common concerning their brains and nervous systems.

What, then, can we make of the comparison between crows and humans? In the main, that the broad definition of intelligence I sketched above is correct. Intelligence is all about the retrieval of information in such a way that its lessons can be generalized and applied to new situations.

Crows and humans, despite their entirely different evolutionary histories, are intelligent in precisely this way. This suggests that intelligence actually means something—something distinct from the parochial conceit of human evolution, and distinct from brain size—and that we should be able to recognize intelligence irrespective of the nature of organism in which it occurs.

If there are intelligent aliens, they needn't, like H. G. Wells's Martians in *The War of the Worlds,* have "intellects vast and cool and unsympathetic," but might be very similar to ourselves.

They'll be liars, cheats, hoodlums, and swindlers.

They'll also be friendly, sociable, sympathetic, and above all, talkative.

And when highly social animals get together, they do like to chat.

10: *The Things We Say*

If heard from a long way off—so you can't hear what's being said—the conversations between parents dropping their children off at primary school probably sound very like the squawks of crows in a wood at evening. The content of both sets of exchanges would probably be very similar, even if you could hear the words spoken by the humans, or make sense of the squawks of the crows.

Is this not a scandalous suggestion? After all, human language—the system whereby we communicate information through the rapid modulation of sounds carried on exhaled packets of air—seems something unique, exceptional. The complicated arrangement of the larynx, the resonant chambers in our nose and mouth, the shape of our throat, the musculature that controls our lips and tongue with great precision and delicacy—all seem refined to a degree seen nowhere else, suitable for conveying the infinite subtleties of language. Apes and monkeys do communicate vocally, but they have nothing to match the sophisticated vocal apparatus of humans, nor, as far as we know, do they indulge in communications of a subtlety that might demand such complexity.

When one looks beyond our immediate primate relatives, we see that many animals have equally unique forms of vocal communication, dependent on equally sophisticated and refined structures. Frogs, for example, display an astonishing range of calls used to attract and respond to prospective mates. Humpback whales have what can only be described as a "culture" of songs, which, like traditional Indian ragas, are built on immensely long sonic structures. Insects of all kinds communicate by chirps, rasps, and buzzes, and the air is full of the songs of birds—created using an organ called the syrinx, every bit as complex and specialized as the larynx, tongue muscles, lips, and so on that we use to create speech. The world is full of the sounds made by animals calling to one another, conveying information.

But what about the complexities of grammar and syntax? Isn't this complexity something that only humans can muster, let alone master? To be sure, human language is rich with meaning and intention. The things we say to one another convey meaning about the ever-changing relationship between people and things in times past, present, and yet to come. To marshal such complexities, the atoms of human language are organized into various categories such as nouns (the names of things) and verbs (which indicate actions and the relationships between nouns), both of which can be modified by adjectives (which modify nouns), adverbs (the same, for verbs), and many other particles that indicate gender, person, time (that is, tense), place, and the relationships between all of these.

Against a pearl of language such as this—

> To be, or not to be, that is the question:
> Whether 'tis nobler in the mind to suffer
> The slings and arrows of outrageous fortune,
> Or to take arms against a sea of troubles,
> And by opposing, end them.

—the bark of a dog seems no more than an involuntary exclamation. But the apparent meaning of words and the relationships between the words, and between their meanings, is just for starters. Human language conveys layer upon layer of *implicit* meaning that can only be understood by the context in which the speech is uttered, and with reference to shared cultural norms.

Figures of speech such as similes and metaphors draw on cultural referents not directly encoded in the text but which will be apparent to the reader, without which the actual words used make no sense. So, when Hamlet talks of slings and arrows, he doesn't mean actual weapons—more the effects of "outrageous fortune."[1] The depth to which cultural convention influences language is a source of much misunderstanding (and humor) when cultures clash. Jared Diamond recalled to me once how he'd got into trouble in New Guinea when he used the pidgin word "pushim" mistakenly to mean "to push" when in pidgin it actually means sexual intercourse. If this seems terribly exotic, think of our own euphemistic sense of the verb "sleep." For example, when Patti LaBelle (in her song "Lady Marmalade") purrs "voulez-vous couchez avec moi ce soir?" she has something more earthy in mind than a slumber party for the kids.

Examples of unintentionally funny translations in public speech and signage are legion—such as the notice in a hotel room inviting guests to "take advantage of the chambermaid"—and the possibly apocryphal tale of how Winston Churchill decided (against advice) to address an audience of Free French in their own language. "Quand je regard mon derrière," boomed Britain's great wartime leader, "je regarde qu'il est divisé en deux parts."

But one doesn't have to look to losses in translation to find humor that takes advantage of the subtlety of language. One of my favorite examples[2] is the newspaper headline from World War II that read

EIGHTH ARMY PUSH BOTTLES UP GERMANS

Indeed, there are words and phrases that, when their cultural referents are taken away, would seem no more meaningful than, say, the bark of a dog, or the clearing of one's throat. Here is one:

If.

As an isolate, this could mean anything. It could be the first word in the eponymous poem by Rudyard Kipling:

If you can keep your head when all about you
Are losing theirs and blaming it on you;

or the last one in a verse in Lewis Carroll's whimsy, Humpty Dumpty's recitation:

He said "I'd go and wake them, if—"

What I am thinking of is the single-word response of the king of Sparta to threats of invasion by Philip II of Macedon when, with all the other Greek city-states having submitted, Philip II advised the Spartans to surrender, having said words to the effect that *if* he invaded Spartan territory he'd kill all the men, violate all the women, enslave all the children, raze the city to the ground, plow salt into the fields so that nothing would ever grow there again, and so on and so forth in a similar bloodthirsty vein. The single-word rejoinder seems hardly more than a grunt, yet it was freighted with the reputation of Sparta, as fierce in battle as spare with words, such that Philip avoided it—as did his son, Alexander the Great. But the simple word "if," when isolated, gives no clue whatsoever about the meaning and its interpretation in context.

To go further, people sometimes say one thing when what they mean is quite different. Steven Pinker gives an excellent example in his book *The Language Instinct*. When asked by a prospective employer to supply a reference for a candidate, the previous employer can hardly say that the candidate is (in Pinker's words) "as dumb as a tree." On the face of it, the reference letter (you'll have to read Pinker's book for the whole example) seems very positive, but on close inspection, it is clear that it offers a very negative report by virtue of the fact that it discusses everything *except* the candidate's suitability.

In this context I might mention an Internet meme called "What Brits Say versus What They Mean," which makes light of the British tendency for reserve, and to avoid embarrassment at all costs. For example, when Brits say that something is "very interesting," they mean that it's "clearly nonsense"; or when Brits say that "it's my fault," they mean that it's *your* fault; and so on.

The language we use is laden with subtlety. But does the fact that we humans use and misuse it without apparent effort make it special? It is not as if one can claim any extra human know-how to be able to use language. As Steven Pinker reminds us in *The Language Instinct*, no human society has ever been discovered that lacks language. The languages of "Stone Age" tribes are as complex, and sometimes much more so, as those of more "developed" societies. But wherever they're from, and irrespective of the culture of the speakers, all language appears to obey the same underlying set of rules, the organization into verbs with tenses, nouns with cases, and so on. It's elegant, it's beautiful, and it's *hardwired*—every bit as the instructions for making hand axes were in the brains of *Homo erectus*, or the instructions for making nests are hardwired into the brains of birds.

But don't we depend on a learning environment in order to translate that hardwired potential into jabbering reality? Isn't the special thing about humans that they learn, rather than operating on instinct? Don't human infants learn to use language only if they are raised in the milieu of older, language-using humans? Yes—but the same is true for other animals that communicate. Humans all have the innate capability of using language, but can only exercise that capability by being raised among speakers of language. Humpback whales can all sing, but they learn their repertoire from other whales.[3] Young male songbirds learn vocal tips from their more experienced elders.[4] So it is that human babies learn from other humans in a similar way, and with the same unconscious, undirected ease. It is perhaps significant that if human children aren't exposed to language during a particular phase of development, they find it very hard to learn later. We all know how hard it is to learn a new language when we're adults. By the same token, dogs that aren't "socialized" through exposure to people and other dogs during a brief "window" of development as pups, may develop as morose, ill-adjusted, and violent.

As a phenomenon, then, language is just one facet of the social behavior of a sociable species. Learning language, like learning to be a sociable member of society, is something we see as human, but the same kind of learning is seen in many social animals that communicate.

That doesn't answer the question, though, of whether human language is either quantitatively or qualitatively more subtle and complex than the systems of communication used by other creatures, such that its possession and use elevates us above all creation. This is a concept with which we all instinctively agree. One of the first things that Adam does in the Bible is give names to the freshly minted animals and plants—this is even before Eve appears (Genesis 2:19–20). Having read this far, however, you'll no doubt appreciate that any assumption of human superiority in this regard will be as suspect as it is in any other.

This assumption—of the superiority of human language, as regards its complexity—relies on an additional, implicit assumption. That is, that it is possible to compare different modes of communication between species and assess them for complexity. The problem with this is that whereas it is possible to measure the raw information content of any signal, such an analysis will not tell us what that signal actually *means*. Moreover, if human language is a trait of humans that is distinctively and uniquely human, it follows that features of communication unique to any given species cannot, by definition, be compared with those of any other, simply because different animals experience the world in different ways. Looked at in this way, it's plain that whale communication is uniquely whale: it probably cannot be rendered simply into human language, and will perhaps be unintelligible to us. There will be aspects of it that we, as humans, will not be able to grasp, simply because of the inherent "whaleness" of its context.

The songs of larks could well mean very much more to other larks than we could ever understand. When you see a male skylark flying high in the sky, the tiny bird producing song of such volume and quantity that you're amazed he doesn't burst, you are sure that he is communicating *something*, else he wouldn't go to all that effort. Your presumption—entirely fair, because it is borne out by the evidence—is that he is singing to attract mates. But that says nothing about precisely what, if anything, he's singing *about*. If he's singing about love and sex, then one could say the same for most human popular music, and quite a lot of unpopular music. If the songs of skylarks have no inherent meaning, in the sense of words and grammar and syntax—one might say the same of much instrumental music, or scat singing in jazz. That the lyrics of "Lady Marmalade" are sexually explicit is undeniable, but music exerts an emotional power even if we can't understand the words, or if there are no words at all. Beethoven's Sixth Symphony—the *Pastoral*—can move me to tears. As a young driver, in my twenties, I would avoid playing heavy rock on the car stereo. Not because I didn't like heavy rock, because I adored it and still do, but because it made me drive more aggressively.[5]

And that's just for species that communicate by sound, as we do. I have not mentioned the subtleties of pheromonal communication in ants, or the waggle dances of bees, or the scent trails of voles. We can get a good idea of what these modes of communications do, in terms of raw function, but of implicit meanings, if any, we will be blind.

So, if the loading of social and cultural context makes the translation of phrases between one human language and another so difficult, imagine how hard it must be to translate languages precisely between different species. To us, the caw of a crow is just that, a caw—but to a crow, that proverbially laconic "if" would seem equally meaningless.

All right then, you might say: even if we concede that there's no *qualitative* difference between the language of humans and the various modes of communication between social animals, don't we humans talk about more elevated things than the matters that (we assume) preoccupy the rest of creation?

No, not really. We do not, as a rule, make idle chat about the tides of politics, or the great unanswered questions of philosophy. Go back to that crowd of gossiping parents in a schoolyard, and listen to what they are talking about. It'll be chat about themselves, their children, their friends, and their everyday social interactions. Many—perhaps most—of the things we talk about can be boiled down to what anthropologists call "social grooming." In his book *Grooming, Gossip and the Evolution of Language*, anthropologist Robin Dunbar argues that language is really about the affirmation of relative social standing. If this seems somewhat harsh, just ask yourself this question. Why, when meeting another person, is it considered polite to inquire about that person's *health*? "How are you?" we ask. Why that, when the world is full of potentially interesting topics of conversation? After all, we *could* kick off a conversation with a complete stranger on practically anything we liked—science policy in Mexico under the government of Carlos Salinas, for example; the problems of rendering the rhyming structure of the Middle English poem *Pearl* into satisfying Modern English stanzas; or the disquieting excess (for the Standard Model of physics) of gamma-ray photons produced in the decay of the purported Higgs boson at the Large Hadron Collider—anything.

But if we did, we'd probably be thought somewhat unhinged and therefore avoided. People who come up to you and start talking about trains are usually regarded as occupying a station on the autism spectrum—a personality trait in which people have trouble responding to social norms.

Comments to strangers that veer away from the conventional how-are-you tactic, yet that are deemed socially acceptable, might be based on a shared sense of identity. For example, if I walk around the streets of the fine city of Norwich while wearing my beanie and scarf emblazoned with the noble emblem and colors of Norwich City Football Club, I am likely to be accosted by another fan, and we'd start a conversation about the ups and downs of our favorite soccer team—and this is a complete stranger, a person with whom I might not have had any other interaction whatsoever. A shared sense of identity sometimes transcends individual recognition. If I wore the same garb in Ipswich, however, I might get beaten up.

Going back to the how-are-you gambit, one might ask in addition why it is considered rude, or at the very least eccentric, if we receive any answer more complicated than a simple affirmation that yes, we're just fine, thank you. In which case, one might ask, why ask the how-are-you question in the first place? Because the question has nothing to do with speech or language at all—its function is to engender social grooming.

Most of the time we don't stop to think about how conventional and ritualized the bulk of human social interaction really is. Language serves to punctuate that interaction, rather than to inform it. That's why it's slightly shocking (and funny) to learn of the habit of a former colleague who, when exhorted by staff in a restaurant or hotel to "have a nice day," would reply, with commanding hauteur—"I have *other plans*." The polite, formalized inquiries we make after peoples' health (or, if in England, to pass some comment about the weather) are no different from the occasions in which dogs stop to sniff each other's bottoms, or baboons stop to pick lice out of each other's fur. Each in its way gathers information about the health of the (for want of a better word) interlocutor.

Most of the rest of what people talk about is gossip about things that happened to themselves and other people: about what she said to him, what he said to her, who did what to whom, what happened next and what it all cost, with many pauses to appreciate the social ups and downs involved: the shame and the schadenfreude. Not that some people don't want to talk about other things: C. S. Lewis, the longtime friend and colleague of the philologist and author J. R. R. Tolkien, once (rather cruelly) remarked that the friends of the relatively uneducated Mrs. Tolkien were the kind of people whose general conversation was "almost wholly narrative."[6] Oh, the irony—this from the doyen of Icelandic sagas. Gossip is, on the face of it, banal. So why do we find it so compelling? As Mozart remarked to his patrons (at least according to playwright Peter Shaffer, in his play *Amadeus*), who wouldn't rather talk with their hairdresser than Hercules? Who, when they should really be doing their homework, or writing this book, wouldn't rather log into Facebook to see what their friends are chatting about? Getting interested in abstract, nonnarrative matters requires a special degree of effort. Gossip, on the other hand, is something we can just fall into and instinctively enjoy.

I think it's fair to say that our love for gossip goes beyond face-to-face interaction, chats on the telephone, and social networks. Most of what people read or hear about in the news or in popular dramas and soap operas (all of which are functionally interchangeable) is social grooming, although at one remove. Think about the human element in any news story you might hear, or read about, or watch on TV, particularly if aimed at a popular audience. Such tales are often about the minor doings of "celebrities"—that is, people who are familiar to us from other contexts, and with whom we identify although we do not know them personally. News grades into "reality" TV, which grades into soap opera. It's all about catastrophe, disaster, human tolls, shock, disgrace, or humiliation—the Shame, and the Schadenfreude. Stories that help one recalibrate one's own position in society. The only thing worse than being talked about, said Oscar Wilde, is *not* being talked about.

How very crow-like we all are.

Does human gossip differ qualitatively, in terms of its elaboration of structure, from that of other animals? To be sure, humans can compose, relate, and understand stories of highly elaborate construction. By this I mean that human stories contain many layers of meaning and action and still remain intelligible to the listener. One can just about follow a sentence such as "Alice thought Bob had told Carol about Donald's invoice to Erica for the work that Fred had done for George," for example, even though it contains four nested stories.

1. Alice thought
2. Bob had told Carol about
3. Donald's invoice to Erica for
4. the work that Fred had done for George.

This nesting is related to what Robin Dunbar calls "intentionality." This relies on our ability to conceive of the mental states of others, but we rely on language to organize it. The sentence above contains four "orders" of intentionality, and Dunbar suggests that human beings are capable of understanding at most six levels of intentionality without having to write everything down or having it explained.

The problem we run into, once again, is that of comparison between species that have very different outlooks on life. Although Dunbar discusses research suggesting that some apes might be capable of third-order intentionality, results can only ever remain that—suggestive. It is hard enough getting into the mind of another animal without having to find reliable ways of discerning whether it is thinking of what another animal is thinking about a third animal, and so on. The question, then, remains open—it is possible that many animals are capable of thinking in this way. And given that most people will not be called upon to understand sentences as complex as this in most situations, one could easily say that there is no functional, real-world difference between the complexities of discourse between animals and between humans.

If language isn't uniquely human, either in its function or its complexity, what about *writing*, the recording of language in symbolic form outside the body, such that it can be preserved and disseminated far more widely than spoken language ever can? Because of writing, we no longer have to learn everything anew in each generation, or rely on oral tradition that disseminates slowly, is prone to error, and can be conveyed to only a few people at once.

Isn't writing—and, by extension, the power of writing to address many people at once—the key to the current domination of the earth by humans? Well, perhaps. Except that many animals use such extracorporeal forms of communication, and many of these animals have little in the way of language, and perhaps less of brains. One thinks of the pheromone trails of ants, or the urine trails of voles—and these are just two examples of extracorporeal communication and reporting that we know about.

A more philosophical problem is recognizing as *representational* any signs or actions made by other creatures. How do we know that the architecture of termite nests isn't random, but a purposeful inscription of their history? This idea might seem outrageous, but a current problem in anthropology is learning how to recognize whether scrawls made by early humans were just inchoate doodles or deliberate records left by inquiring minds.[7] And if such things are hard to judge for members of our own species, we can have little hope that we might recognize, still less decipher, any form of extracorporeal communication left by other animals.

I contend, however, that at least some extracorporeal forms of communication are just that—representational—in that they contain particular meanings that are there to be interpreted by others of the same species once the author has left the scene. The example is, however, personal and anecdotal—because I have personal experience of it on a daily basis—and that is the intensely odoriferous imagination of dogs and the ability of dogs to leave messages to be interpreted by other dogs without direct dog-to-dog contact.

Most days when I take my dogs out walking, it's not the exercise they seem to crave, but the social stimulation. An invitation to go for a walk is greeted with intense excitement—much barking, wagging of tails, and general jumping up and down—but a gentle amble of less than two miles leaves them completely exhausted. It's the social stimulation, I think, that saps them—not the actual physical exercise. Every few steps we stop so that the dogs can sniff what seem to them to be interesting blades of grass, lampposts, walls, tree stumps, and so on. They sniff with the deliberation of master wine tasters, and, sometimes, mark the site with small urine samples of their own. To me, the human observer, it looks just as if they are sampling the status updates of other dogs on their doggy social network—let's call it SniffBook—and perhaps leaving their own comments. We humans have a very poor sense of smell compared with that of dogs, so we cannot appreciate the refinement, the nuance, the bouquet—the *meaning*—of the scents the dogs are exchanging, but given what we know about gossip in general, and the importance of social networks, we can have a good guess. The dogs are trading information about who's who, who's been around, who has said what to whom, and, perhaps most of all, *their state of health*. We humans do it through vision, language, and sound—dogs do it through smell. The modality is different, but the end result is just the same.

I live on the very picturesque coast of north Norfolk in England, which is great dog-walking country. Being proverbially flat, it's perfect for an easy ramble. The beaches are vast and deserted; the skies are enormous and dramatic. One morning my wife had arranged to meet with two friends for a walk, and she asked me to drive her and our dogs to a cliff-top rendezvous, whence they'd make their own way home. Her friends were there, waiting, with their own dogs, as we arrived. As I drove away, I was much taken by the scene in the rearview mirror. A meter and a half above ground level, the humans were talking animatedly with one another, mouths moving, hands waving—a meter below that, the dogs were greeting one another with equal enthusiasm, with much sniffing of bottoms and wagging of tails. Without being distracted by the words uttered by the humans, it seemed to me that the behavior of humans and dogs was all but identical irrespective of the mode of communication. I felt like the animals at the end of George Orwell's *Animal Farm*, peering through the window as the pigs and the farmers feast convivially inside, looking first at the pigs, and then at the farmers, and at the pigs again, and finally not being able to decide who was who.

11: *The Way We Think*

So much for bipedality. So much for large brains, technology, intelligence, and language. There might—just might—be one ability, one trait, that marks us out as special. We human beings, surely, differ from other animals in that we are conscious.[1] That is, we are aware of our actions and their consequences, having the ability to imagine ourselves as participants in the drama of our own lives, and how our lives interact with those of others.

I find the term "consciousness" rather vague, and so the effort to understand it is as challenging as trying to nail jelly to the ceiling. I prefer "self-awareness," the meaning of which is self-explanatory: that we have a sense of "self," as if we are a whole, cohesive entity, inhabiting a body. In this book I use the term "sentience" rather than "self-awareness," because it is shorter and more elegant, but my meaning is precisely the same. A sentient being will be aware of itself as a character in the drama of its life, and thus aware of the consequences of its actions on others, and to some extent of the internal mental states of the other characters. Psychologists might say that sentient beings have "a theory of mind."

Art, religion, even science, spring directly from this sense of self. Sentience brings along with it the crushing realization of the brevity of life, the inevitability of death, and through that, a desire to investigate and explain the human condition.

The poet John Keats knew all about this, perhaps better than anyone. As a young apothecary in early nineteenth-century England, the business of disease, debilitation, disfigurement, and death was part of his daily round. He saw his relatives and friends sicken and die young, mainly from tuberculosis, to which he, too, eventually succumbed. The transience of life was well expressed in his epitaph: "Here lies one whose name was writ in water." Yet in a desperately short life—he died before he was twenty-six—he created arguably the greatest poetry ever written in English.

In his great poem *Ode to a Nightingale* (written in May 1819), he contrasts the pain of a mortal doomed to muse on his lot, with the joy of the nightingale, living ever for the moment and therefore not doomed to death, a concept that would mean nothing to the insentient.

> Fade far away, dissolve, and quite forget
> What thou among the leaves has never known,
> The weariness, the fever and the fret
> Here, where men sit and hear each other groan;
> Where palsy shakes a few, sad, last gray hairs,
> Where youth grows pale, and spectre-thin, and dies,
> Where but to think is to be full of sorrow
> And leaden-eyed despairs;
> Where Beauty cannot keep her lustrous eyes,
> Or new Love pine at them beyond to-morrow.

But sentience—the luxury of self-knowledge—is in fact not unique to humans, and its presence in other animals can be tested and verified. Clayton and her colleagues have shown that crows modify their behavior in predictable ways depending on the identity, number, and attitudes of other crows in the vicinity.[2]

In one experiment, a western scrub jay buries some food in full view of other jays, but will then return when no other jays are around, unearth the cache, and bury it somewhere else. Significantly, the jays that rebury their food in private are those that had been thieves of the caches of others in the past. The conclusion seems clear—the jay pictures itself as a participant in a drama in which it guesses the intentions of other jays close by, which would be to steal its food store. The jay seems to be able to put itself in the minds of its fellows, imagining what it would do in a similar situation.[3] Sentience is a valuable asset for any social animal, but with sentience comes deceit. It is probably no coincidence that very young children are very bad liars until around the age of three, when they first acquire a "theory of mind" and can put themselves in the shoes of others. By the same token, people with autism-spectrum disorders can have great difficulty in social situations because they have trouble reading the emotional states of others,[4] and must learn by intellection what others seem to absorb by instinct. Autism-spectrum disorders might therefore be seen as disorders of sentience.

Sentience, however, is not an unalloyed benefit . . .

Hold it right there: how can something that seems so beneficial, so wonderful, that it might easily be seen as the final attainment of humanity, the justification of an exalted place as the acme and purpose of Creation, the final revelatory light that switches on in our minds, from which flows all art, culture, science, and indeed everything that seems to make human life so much richer and more distinctive from that of any other organism, be seen as in any way a disadvantage?

Well, let's start with something we all know, and some of us remember with much toe-curling embarrassment: our teenage years. One might interpret the extreme self-consciousness of teenagers, whose brains are undergoing drastic remodeling before the final attainment of adulthood,[5] as a disarming and sometimes crippling excess of sentience. Teenagers try to grapple, perhaps for the first time in their lives, with age-old questions—questions such as the meaning of existence, man's inhumanity to man, and so on—that their parents have long abandoned in favor of more tractable problems, such as the location of one's spectacles, or the identity of whoever it was that put the Benzedrine in Mrs. Murphy's Ovaltine.[6] It's perhaps no accident that the greatest artists, poets, musicians, mathematicians, and even scientists tend to do their best work when they are young, when their self-knowledge is at its peak. To paraphrase Tom Lehrer: when Mozart was my age, he'd been dead for fifteen years.

Everyone who's been a teenager will have experienced the same agonies of self-consciousness, and will have been relieved, frankly, when that fit's over. But if teenage sentience can be a handicap, just imagine how difficult life would be, intolerable even, if we were sentient *all the time*.

When one is learning a new skill, whether it's a sport, driving a car, or learning a musical instrument, one is often painfully aware of every movement one makes, and wonders if one will ever get the hang of it. With practice, however, the movements we make as we perform these tasks become automatic, wired into those parts of the brain that do things without our having to be conscious of them.

That's why an experienced driver, say, will be able to drive along a familiar route literally without thinking about every turn of the wheel, every press on the gas or the brake, and will be able to take evasive action (such as swerving out of the path of an oncoming vehicle, or applying the brakes before a potential collision) faster than conscious thought would seem to allow. When a driver finishes his journey, he will not be able to recall the precise sequence of actions he took as he drove, as he would were he a computer. A concert pianist will be able to play a complex, learned piece by letting her fingers do the walking with what musicians call "muscle memory," using the sheet music only as a backup.

I believe that we live most of our lives in this way. Just as we don't give conscious thoughts to routine, learned habits such as driving, we do not, as a rule, record in any self-conscious way the moments of our lives as they pass. When you look back at one day lived, you recall a small series of incidents as blurry snapshots, not every single moment as you lived it in exhaustive detail. The vast bulk of the time through which we travel is passed in a state of insentience.

In fact, I'd go so far as saying that most people live most of their lives without much being troubled by sentience. Is this not a somewhat snobbish attitude? To be sure, you could see it that way, but consider the alternative—and if you do, you'll see that it is almost too horrible to contemplate.

The Argentine essayist Jorge Luis Borges did just that in a memorable story called *Funes the Memorious*.[7] Ireneo Funes is a young man who, as a result of a head injury, has perfect recall of every moment of his life. The effect is disabling: because he sees and records in perfect detail, he can no longer categorize objects, for every new thing he sees is unique. For example, Funes is unable to recognize any individual dog as a member of a class of creature called "dogs," by abstracting those features that all dogs have in common. Funes would have read as meaningless Ogden Nash's prescription in *Introduction to Dogs*:

> The dog is man's best friend.
> He has a tail on one end.
> Up in front he has teeth.
> And four legs underneath.

Because Funes sees every detail of every dog, he is unable to distinguish between those features that are specific to each dog, and those that belong to dogs more generally. To Funes, each dog is sui generis: so distinct, one from the other, that no categorization is possible.

By the same token, Funes remembers everything that happens with perfect clarity, and is therefore unable to summarize any one day of his life by abstracting any highlights—to us, the snapshots we remember—as in doing so he is forced to relive each day in real time. Everything in his life is important, and, as a result, nothing is. Incapable of judgment, he is confined to a single room, paralyzed by self-awareness. It seems clear that while sentience has adaptive value for social creatures, one might have too much of a good thing.

All of the above rests on a single, untested assumption about the mechanics of sentience. To be sentient—to have a "theory of mind"—you must be able to imagine yourself in the drama of your own life, as if you were an actor on a stage along with imagined representations of your friends and relations.

Now, here's the thing—if you're all on this imaginary stage strutting your stuff, *who is the audience*? The conventional answer is that you yourself are the audience. But to picture *that*, you have in a sense to be watching yourself watch the drama, in which case there has to be another version of you watching the watcher, and yet another watching the watcher of the watcher, and so on—an infinite hall of mirrors. In the mind's theater the watchers come and go, toward *absurdam, reductio*.

In *Consciousness Explained*, philosopher Daniel Dennett shows that this image of a mental theater might make a pleasing metaphor, but it is almost certainly not how the mind works. The philosopher René Descartes imagined that the "soul," or in our terms our sense of "self," was located in a physical part of the brain (he chose the pineal gland), but no evidence has ever come to light that any physical part of the brain corresponds with this so-called Cartesian theater. There is no central command center, like the bridge of a ship through which lots of little people look out through our eyes as windows, surveying the world and acting on information received. In terms of actual anatomy, rather than metaphor, there is no single part of the brain that processes all incoming sensory information, integrates it, mulls it over, and then instructs the appropriate responses.

Sensory information does come in, and is processed by various parts of the brain—but the processing is piecemeal, done by several different parts of the brain. Eventually, responses are formed, but again, not in any straightforward way that depends on the inputs. Indeed, the assumption even by trained neuroscientists and philosophers that there must, somewhere, be something akin to the Cartesian theater has led to all sorts of seemingly anomalous research results, perhaps most notably the initiation of a motor action before the subject is "conscious" of taking that action—a result that has led to all sorts of questions about free will.[8] The simpler solution—but somehow the harder one to take—is that there is no single center of consciousness. There is no Cartesian theater, no command center, no captain's bridge. Sentience is an illusion, a kind of running commentary kludged together after the fact, by and for the benefit of lots of different parts of the brain at once. And the brain is easily fooled.

I am sure you've had dreams in which you are involved in epic dramas, dreams with plots that seem to take a great deal of time to unfold, but that end with some tumultuous noise. You awake and find that the noise is your alarm clock. As you stir yourself into wakefulness, you will naturally ask yourself how your brain laid out such a complex drama so that it culminated *precisely* at the moment your alarm went off. You might say that as you know very well that your alarm is going to go off at (say) seven o'clock—so well that you often wake up at 6:59, just in time to switch it off—then your mental impresario will have started a well-timed sequence of events designed to culminate at that moment.

But that must be wrong, as there have been other occasions when long, complex dreams have ended with some disturbance caused by a sudden, external stimulus that could not have been predicted. Is your mind a time traveler? Can you see the future? Of course not—your senses respond to the stimulus, but your mind makes sense of it backward, rationalizing it after the fact, giving the illusion of the forward passage of time.

We are visual creatures, so it's not surprising that vision has long been the playground of neurobiologists seeking to understand consciousness.[9] Light impinges on the retina, causes an electrochemical change in the optic nerve, creating a signal that travels along the optic nerve to the brain, eventually arriving at the part of the cerebral cortex where the signal is processed. Does this raw signal arrive at an imagined version of a movie screening room? Apparently not—the signal is processed in various ways before it gets to the cerebral cortex. What we "think" we "see" is very far from the pattern of photons that hit the retina. The "image" has been cross-referenced with other sensory data and memories of past images, adding meaning. If this weren't the case, that thieving scrub jay wouldn't be able to tell the difference between its fellow jays and any other bunch of photons. This idea—that nonvisual inputs give meaning to a visual image—is a bugbear for researchers into artificial intelligence, who have difficulty getting a computer to recognize that when an object goes behind another object and comes out the other side, the two are in fact the same, coherent object.

What we see of the world around us is far from a detailed panorama in which everything is in focus, up and down, and from side to side, as if our eyes were cameras. We pay attention only to a small area at any time. If this seems counterintuitive, go take a look at any landscape by the photographer Ansel Adams, and ask yourself why, for all its realism, it looks weird and dreamlike. The reason is that *everything* is in focus, from the rocks in the foreground to the mountaintops in the distance. We don't actually see the world like that.

Color is likewise tricky. Is there a fundamental quality of, say, "red" that's "out there" in the world? Why do we interpret electromagnetic radiation in a particular range of frequencies as "red"? When you imagine a red sports car, how "red" is it in your mental picture of a sports car? If I hadn't asked you to imagine it, would you have seen it as "red" or—perhaps—as a label you might find in a paint-by-numbers kit, that stands for "red" but isn't actually "red" itself? If this "label" isn't "red," what color is it? Does it have any color at all? If it does, would you have ascribed a color to it had I not inquired about it?

Dennett argues that there isn't any objective reality to "red," in the same way that there isn't a central command unit of the mind. What *is* there, instead, is a reaction that's been honed by natural selection in which the visual system pays especial attention to the electromagnetic radiation in the particular frequency range that is characteristic of ripe fruit when seen against a background of green leaves,[10] or, as it may be—and not uncoincidentally—the swollen genitalia of potentially receptive mates. Whether we call it "red," "borogove," or "manxome" is immaterial, because the explanation does not require the postulation of any kind of self-conscious inner life.

Sentience, then, is a slippery customer. One might be tempted—I *am* so tempted—to say that it doesn't exist—not, at least, as a discrete quality.

But if that is the case, how can we interpret the behavior of Clayton's scrub jays, or indeed of any human or animal that shows a "theory of mind" or various degrees of what Robin Dunbar calls "intentionality"?

I think that whereas such behavior *appears* to show that creatures have a sense of what is going on in the minds of others, that appearance is illusory, conditioned by our own preconceptions about sentience—that it is discrete, and, moreover, that it happens in a Cartesian theater. To put it another way, we interpret the scrub jays' behavior for what it is because we recognize that behavior in ourselves, and we attribute that behavior to a "theory of mind." We therefore project that theory into the heads of jays, rather in the way that our minds make sense, retrospectively, of things that go bump in the night.

What the scrub jays are *actually doing* is exhibiting a behavior that is selectively advantageous in a social situation. Once that is realized, the necessity for a quality called "sentience" becomes moot, for natural selection does all sorts of wonderful things, produces all sorts of exquisite adaptations that, when studied casually, have all the appearance of intention and design, yet we know that nothing of the kind has taken place. In this light we can perhaps see the idea of sentience as the last great conceit of that linear view of evolutionary progress that has humanity marching at its head, rather than of Darwin's original concept of a tangled bank upon which circumstances changed from one instant to the next with no end in view, no purpose to attain. The tangled bank might, indeed, make a better metaphor for the workings of the mind than the Cartesian theater—so clean, so tidy, and so *wrong*.

If there is no Cartesian theater, what becomes of the "self" that both watches and takes part? Does the "self" really exist? If there is no "self," you might ask, whose arthritis is this? In my view the self is as illusory as sentience and the Cartesian theater. To reprise that old limerick:

> There was a faith-healer of Deal
> Who said, "I know that pain isn't real:
> But when I sit on a pin
> And it punctures my skin,
> I dislike what I *fancy* I feel."

The idea of a "self" is not only illusory—it's unnecessary. There need be no "inner life" if such traits can be judged in the context of externals, of observable behavior and social interaction. There is an area of research into animal behavior in which it is shown that animals of the same species may have measurably different "personalities."[11] The idea of "personality" in this context is deliberately quite narrow. A "personality" means a self-consistent set of responses that makes a discrete behavioral repertoire or "syndrome." For example, some animals tend to be more outgoing than others—bolder and more risk taking than their fellows, who might be shyer and more risk averse. Anecdotally, anyone who owns a dog will be aware that pets have personalities. My golden retriever, Heidi, is phlegmatic and laid-back, whereas my Jack Russell terrier, Saffron, is excitable, a real live wire. Domestic dogs belonging to distinct breeds are, however, highly inbred, and their temperaments are to some degree breed-specific. Some dog breeds are prone to behavior that parallels various psychiatric problems in humans, such as anxiety or obsessive-compulsive disorder: the fact that dogs are inbred compared with humans is allowing researchers to get a fix on the genetic roots, if any, of such traits, in the hope that the human cognates of such genes might be identified.[12]

None of this bears on whether dogs are sentient. Strictly, animal personalities reflect consistent behavioral differences of possible selective value[13]—researchers in this field scrupulously avoid the philosophical quagmires of sentience and theories of mind. In which case, it is perfectly possible to view the behavior of apparently sentient creatures such as crows and people in this same dispassionate manner—as it is the behavior of creatures that have little or nothing in the way of brains at all.

Aha, I hear you cry—there is an observable way to discern whether an animal has a sense of "self," and that's by mirror self-recognition. That is, whether an animal can look at its reflection in a mirror and recognize it as a representation of itself rather than another animal, or nothing at all. Cats and dogs don't have this sense. When they look in a mirror, they definitely see *something*, and by the aggressive attitude they sometimes adopt, it is likely to be another cat or dog. It doesn't see its "self" in its reflection.

The classic test for mirror self-recognition is the "mark test." The experimenter marks an animal in such a way that it cannot see the mark except in its reflection. If, on seeing the reflected mark, the animal looks for it on its own body, it is therefore able to recognize its reflection as itself rather than another animal or anything else. Humans are able to do this from a very early age,[14] and the mark test has been passed by the great apes, elephants, dolphins, and—you'll not be surprised to learn—at least one member of the crow family.[15]

By now you'll have spotted the trapdoor lurking underneath such research. If you haven't, it's this—as with theories of mind, the results of such tests are biased by our own capabilities and expectations. Being, as we are, visual creatures, who recognize one another largely by visual inspection, we tend to judge other creatures by the same yardstick, and unconsciously rank them according to the age-old ladder of creation in which humans are at the top. We have no idea whether creatures are able to recognize themselves using other senses, such as smell. Given our ignorance of the olfactory universe of dogs, for example, we simply don't know whether dogs can identify their own smells. However, observing one's dogs as they sniff lampposts and one another's bottoms, one suspects that at the very least dogs are able to recognize and distinguish between other dogs, and possibly between different humans.

Is there any experimental evidence for self-nonself recognition that is not based on vision? Yes—plenty of it. For example, it is known that quite a few animals are able to distinguish between kin and nonkin,[16] presumably through olfactory cues in urine. In mammals, at least, this recognition is based on a region of the genome called the major histocompatibility complex, or MHC. Breakdown products of MHC proteins are excreted in the urine and can be detected by creatures of sufficiently sensitive sniffage. The MHC is part of the immune system—it produces a set of molecular markers of incredible variety that identify any matter in the bloodstream that comes from another organism, such as an infectious agent. It is this system that makes matching organ donors with recipients so difficult. The same system is exploited by animals, including social animals, to tell the difference between close relations and nonkin, perhaps as a way to avoid inbreeding. The MHC is really just a riff on a very basic self-nonself recognition system seen in many creatures, even some quite lowly ones.

The ability to recognize oneself in a mirror is, therefore, a somewhat contrived special case of a much more general ability to distinguish between oneself and other creatures, based on senses that aren't necessarily visual. If one thinks of mirror self-recognition as a good marker of sentience, try this thought experiment. Would we humans, with our notoriously poor sense of smell, be able to distinguish between ourselves and other humans on the basis of smell—from urine samples, body odor, or sniffing or licking one another's' bottoms or genitalia—with the same reliability as we can tell the difference between people by sight?

If not, why should we expect other creatures, perhaps with visual capabilities that aren't as good as ours, to recognize one another by sight alone?

Sentience, that seemingly quintessential human characteristic, is an artifact maintained by human exceptionalism—and has given up with hardly a whimper.

Afterword: The Tangled Bank

Several people who read excerpts of the draft of this book have, understandably, been distressed that I should walk away from this extended dustup without leaving even one tiny crumb of comfort. What people want, I guess, is the conventional Hollywood happy ending. In the movies, the aliens never destroy the earth and disappear, leaving a charred ruin. No, the human underdogs fight back against overwhelming odds, employing well-known heroic virtues such as motherhood and apple pie, and drive the alien insectoid scum to destruction.[1] It's a tale as old as time, to coin a phrase: as old as David vs. Goliath, Frodo vs. Sauron, Luke vs. Vader, Harry vs. Voldemort.

We human beings do like to tell stories, and the conventional picture of evolution as a stately and predictable procession with humanity at its head is just that—a story. As such, it speaks both to a profound misreading of Darwinian evolution, and to assumptions based on the fossil record that it cannot support, and never will.[2]

Appeals to various human traits sometimes regarded as qualitatively "special" likewise fail. This is because the various traits expressed by organisms, human beings included, don't represent way stations between Ape and Angel, or even one species and another, but contingent, makeshift compromises made in response to a number of different factors, some reinforcing one another, others acting in opposition. Because of this, such traits cannot be treated in isolation. Organisms don't evolve first one trait and then another, hauling themselves up the ladder of creation at each step, but in an integrated way, as you'd expect for creatures living on the tangled bank of Darwin's imagination.

Bipedalism, for example, didn't evolve, on its own, "in order that" we might free our hands to make tools, share food, or any other purpose conferred after the fact. To be sure, such factors might have been involved in the transition, along with sexual selection, the evolution of a social life, the invention of cookery, the growth of the brain, the dependency of infants, the further growth of social groups, the invention of technology, the evolution of the menopause, the evolution of language, the telling of tales, and so on. Bipedalism required a wholesale change in hominin anatomy, yet, considered as a whole, was just one change among many that happened in the hominin lineage, a change that facilitated (or impeded) changes elsewhere in the body, and was in turn influenced by these changes.

Even after the foregoing, one might argue that humans aren't qualitatively special for what they are, nor even for how they got that way, but for what they've achieved. One might well ask, for example, whether humans are unique in that they have a conception of the divine, the numinous. It's impossible to know. The gospels of crows, the dreams of dolphins, and the speculations of anthills might be forever beyond our grasp.

Books of this kind usually end by looking forward to humanity's glorious future in space (if optimistic), humanity's capacity for destructive warfare (if pessimistic), or humanity's destruction of the environment (ditto). Humans aren't unique in any of these achievements, either. Bacteria can survive exposure to space,[3] even without space suits, and there seems no good reason why, if living organisms can withstand hard radiation, they cannot survive routinely in the uppermost atmosphere. Ants are accomplished warriors. And any effects we might have on the environment, while regrettable in themselves, are puny compared with the effects of the bacteria that, billions of years ago, invented a nifty scheme for harvesting sunlight to drive a process in which plentiful carbon dioxide and water were made into food. This process, oxygenic photosynthesis, changed the world utterly, not least because its by-product, molecular oxygen, was and is both toxic and corrosive. The emission of large amounts of oxygen into the atmosphere drove much of the then-existing biosphere to the brink of extinction and rotted the very bones of the earth. We are the by-products of the terrific selective forces for creatures that could withstand this assault, and even use it to their advantage.

All such scenarios, though, are told as stories—we cannot help it. They demand narrative, characterization, and a moral payoff, even if the action happened billions of years ago and the protagonists were bacteria. Stories are something we humans can't live without. They fuel our most disposable gossip and inform our deepest speculations. Perhaps, then, the capacity to tell and appreciate stories is the one thing that marks human beings from the crowd.

And they all lived happily ever after.

Or did they?

Notes

PREFACE

1 J. L. Franzen et al., "Complete primate skeleton from the Middle Eocene of Messel in Germany: Morphology and paleobiology," *PLOS One* 4 (2009): e5723, doi:10.1371/journal.pone.0005723.

2 See my essay "Aspirational thinking," *Nature* 420 (2002): 611.

3 H. Gee, "How humans behaved before they behaved like humans," *London Review of Books*, 21 October 1996, 36–37. For the record, the books I reviewed were C. Stringer and R. McKie, *African Exodus: The Origins of Modern Humanity* (London: Cape, 1996); R. Foley, *Humans before Humanity* (London: Blackwell, 1995); C. Tudge, *The Day before Yesterday: Five Million Years of Human History* (London: Cape, 1996); A. Walker and P. Shipman, *The Wisdom of Bones: In Search of Human Origins* (London: Weidenfeld, 1996); and J. Shreeve, *The Neanderthal Enigma: Solving the Mystery of Modern Human Origins* (London: Viking, 1996). But wait, there's more—two years earlier in the same magazine I'd reviewed three others (H. Gee, "What's our line?," *London Review of Books*, 27 January 1994, 19), and they were E. Trinkaus and P. Shipman, *The Neandertals: Changing the Image of Mankind* (London: Cape, 1993); C. Stringer and C. Gamble, *In Search of the Neanderthals: Solving the Puzzle of Human Origins* (London: Thames and Hudson, 1993); and J. Kingdon, *Self-Made Man and His Undoing* (New York: Simon and Schuster, 1993).

4 J. Maddox, *What Remains to Be Discovered: Mapping the Secrets of the Universe, the Origins of Life, and the Future of the Human Race* (New York: Touchstone, 1998).

5 P. Pearson, book review, *Palaeontology Newsletter* 44 (2000): 36–38.

6 "When I was a child, I spake as a child, I understood as a child, I thought as a child: but when I became a man, I put away childish things." 1 Corinthians 13:11.

CHAPTER 1

1 H. Gee, "UK nuclear operators unhappy with radiation dose advice," *Nature* 330 (1987): 596.

2 F. Gibson et al., "A type VII myosin encoded by the mouse deafness gene *shaker-1*," *Nature* 374 (1995): 62–64.

3 R. A. Lutz and J. R. Voight, "Close encounter in the deep," *Nature* 371 (1994): 563.

4 P. Brown et al., "A new small-bodied hominin from the Late Pleistocene of Flores, Indonesia," *Nature* 432 (2004): 1055–1061.

5 See for example S. O'Connor, "New evidence from East Timor contributes to our understanding of earliest modern human colonization east of the Sunda Shelf," *Antiquity* 81 (2007): 523–535; and B. David et al., "Sediment mixing at Nonda Rock: Investigations of stratigraphic integrity at an early archaeological site in northern Australia and implications for human colonization of the continent," *Journal of Quaternary Science* 22 (2007): 449–479.

6 A. Brumm et al., "Hominins on Flores, Indonesia, by one million years ago," *Nature* 464 (2010): 748–752.

7 The term "hominin" might be unfamiliar and deserves a note of explanation. It means any species, living or extinct, that is more closely related to modern humans (*Homo sapiens*) than to our closest living relatives, the chimpanzees (*Pan troglodytes* and *Pan paniscus*). Importantly, this designation does not demand or imply direct ancestry, although it does not rule it out. The most we can say of any hominin other than *Homo sapiens* is that we are related to it, at the very least as cousins. I discuss this concept of relatedness and much else besides in my book *In Search of Deep Time*.

8 M. J. Morwood et al., "Archaeology and age of a new hominin from Flores in eastern Indonesia," *Nature* 431 (2004): 1087–1091.

9 M. R. Palombo, "Endemic elephants of the Mediterranean islands: Knowledge, problems and perspectives," in G. Cavarretta, ed., *The World of Elephants—International Congress* (Rome: CNR, 2001), 486–491.

10 E. M. Northcote, "Size, form and habit of the extinct Maltese swan *Cygnus falconeri*," *Ibis* 124 (1982): 148–158.

11 S. L. Vartanyan et al., "Holocene dwarf mammoths from Wrangel Island in the Siberian Arctic," *Nature* 362 (1993): 337–340.

12 M. J. Morwood et al., "Further evidence for small-bodied hominins from the Late Pleistocene of Flores, Indonesia," *Nature* 437 (2005): 1012–1017.

13 Dean Falk summarizes the story very well in her book *The Fossil Chronicles: How Two Controversial Discoveries Changed Our View of Human Evolution* (Berkeley: University of California Press, 2011). This book is especially interesting given that she has been an important player in the debate. Her work comparing the skulls and brains of humans (both normal and microcephalic), *Homo floresiensis*, and other extinct hominins has led her to conclude that *Homo floresiensis* represents a distinct species rather than a variant of *Homo sapiens*. See for example D. Falk et al., "Brain shape in human microcephalics and *Homo floresiensis*," *Proceedings of the National Academy of Sciences of the USA* 104 (2007): 2513–2518; and D. Falk et al., "The brain of

LB1, *Homo floresiensis*," *Science* 308 (2005): 242–245. More support comes from D. Argue et al., "*Homo floresiensis*: Microcephalic, pygmoid, *Australopithecus*, or *Homo*?," *Journal of Human Evolution* 51 (2006): 360–374.

14 See for example I. Hershkovitz et al., "Comparative skeletal features between *Homo floresiensis* and patients with primary growth hormone insensitivity (Laron syndrome)," *American Journal of Physical Anthropology* 134 (2007): 198–208; G. D. Richards, "Genetic, physiologic and ecogeographic factors contributing to variation in *Homo sapiens*: *Homo floresiensis* reconsidered," *Journal of Evolutionary Biology* 19 (2006): 1744–1767; T. Jacob et al., "Pygmoid Australomelanesian *Homo sapiens* skeletal remains from Liang Bua, Flores: Population affinities and pathological abnormalities," *Proceedings of the National Academy of Sciences of the USA* 103 (2006): 13421–13426.

15 J. Krause et al., "The complete mitochondrial DNA genome of an unknown hominin from southern Siberia," *Nature* 464 (2010): 894–897.

16 R. E. Green et al., "A draft sequence of the Neandertal genome," *Science* 328 (2010): 710–722.

17 See Richard Fortey's book *Survivors: The Animals and Plants That Time Has left Behind* (London: HarperCollins, 2011) for a charming roundup of living fossils.

18 C. Oxnard et al., "Post-cranial skeletons of hypothyroid cretins show a similar anatomical mosaic as *Homo floresiensis*," *PLOS One* 5 (2010): e13018, doi:10.1371/journal.pone.0013018.

19 S. G. Larson et al., "Descriptions of the upper limb skeleton of *Homo floresiensis*," *Journal of Human Evolution* 57 (2009): 555–570; M. W. Tocheri et al., "The primitive wrist of *Homo floresiensis* and its implications for hominin evolution," *Science* 317 (2007): 1743–1745; W. L. Jungers et al., "The foot of *Homo floresiensis*," *Nature* 459 (2009): 81–84.

20 M. M. Lahr and R. Foley, "Human evolution writ small," *Nature* 431 (2004): 1043–1044.

21 D. Lordkipanidze et al., "Postcranial evidence from early *Homo* from Dmanisi, Georgia," *Nature* 449 (2007): 305–310.

22 E. M. Weston and A. M. Lister, "Insular dwarfism in hippos and a model for brain size reduction in *Homo floresiensis*," *Nature* 459 (2009): 85–88.

23 A. D. Gordon et al., "The *Homo floresiensis* cranium (LB1): Size, scaling and early *Homo* affinities," *Proceedings of the National Academy of Sciences of the USA* 105 (2008): 4650–4655.

24 I have heard that the voice-over was provided by a prominent evolutionary biologist, though I should emphasize that this is just gossip.

25 This is a reasonable assumption given that all the great apes—the chimps, the gorillas, and the orangutans—are much more similar to one another, as regards their general features, biology, and life habits, than any one in particular resembles humans. It used to be the case that the apes were placed in a single family, the Pongidae. However, it turns out that chimps

are in fact more closely related to humans than are gorillas, with the orangutans at a more remote remove. This means that the term "Pongidae" doesn't refer to what zoologists call a "natural group"—unless humans are included.

26 This doesn't mean that the common ancestor of all creatures living today was the first creature to have evolved. It is possible that other essays in self-reproducing systems appeared, but became extinct and left no trace. Scientists therefore sometimes refer to the common ancestor of all extant life as LUCA, which stands for "last universal common ancestor."

27 D. L. Theobald, "A formal test of the theory of universal common ancestry," *Nature* 465 (2010): 219–222.

28 R. D. Martin, "Primate origins: Plugging the gaps," *Nature* 363 (1993): 223–234.

29 This is why the expression "this research raises more questions than it answers" is a cliché.

30 P. J. Turnbaugh et al., "The Human Microbiome Project," *Nature* 449 (2007): 804–810.

CHAPTER 2

1 National Health Statistics Report 10, 22 October 2008.

2 http://en.wikipedia.org/wiki/Human_height, accessed 4 January 2013.

3 B. C. Msamati and P. S. Igbigbi, "Anthropometric profile of urban adult black Malawians," *East African Medical Journal* 77 (2000): 364–368.

4 UK National Health Service, http://www.ic.nhs.uk/statistics-and-data-collections/health-and-lifestyles-related-surveys/health-survey-for-england/health-survey-for-england-2008-trend-tables, accessed 22 January 2011.

5 J. Cohen, "Knife-edge of design," *Nature* 411 (2001): 529.

6 I discuss this idea in my book *Jacob's Ladder*.

7 *OED*, June 2011, http://www.oed.com/view/Entry/65447, accessed 6 September 2011. All the examples of the use of the word "evolution" in this chapter come from that source, unless otherwise stated.

8 W. Bateson, *Materials for the Study of Variation, Treated with Especial Regard to Discontinuity in the Origin of Species* (London: Macmillan, 1894), v–vi.

9 L. Oken, *Abriss des Systems der Biologie* (1805); this translation in R. J. Richards, *The Meaning of Evolution* (Chicago: University of Chicago Press, 1992), 39.

10 Most notably William Bateson, a cofounder of the science of genetics.

11 J. Z. Young, *The Life of Vertebrates* (Oxford: Clarendon Press, 1981).

12 N. Eldredge and S. J. Gould, "Punctuated equilibria: An alternative to phyletic gradualism," in T. J. M. Schopf, ed., *Models in Paleobiology* (San Francisco: Freeman, Cooper, 1972), 82–115.

CHAPTER 3

1 For example, see K. Chong, "*Bacillus cereus* in 'Poon Choi,'" in Food Safety Focus (Hong Kong: Centre for Food Safety), http://www.cfs.gov.hk/english/multimedia/multimedia_pub/multimedia_pub_fsf_40_01.html, accessed 29 March 2012.

2 "The hidden hordes," *Nature Reviews Microbiology* 10 (2010): 517.

3 L. Rothschild and R. Mancinelli, "Life in extreme environments," *Nature* 409 (2002): 1092–1101.

4 See for example D. W. Griffin, "Terrestrial microorganisms at an altitude of 20,000 m in Earth's atmosphere," *Aerobiologia* 20 (2004): 135–140; A. P. Teske, "The deep subsurface biosphere is alive and well," *Trends in Microbiology* 13 (2005): 402–404.

5 See for example R. C. Blake II et al., "Chemical transformation of toxic metals by a *Pseudomonas* strain from a toxic waste site," *Environmental Toxicology and Chemistry* 12 (1993): 1365–1376; J. Lloyd and J. C. Renshaw, "Bioremediation of radioactive waste: Radionuclide-microbe interactions in laboratory and field-scale studies," *Current Opinion in Biotechnology* 16 (2005): 254–260.

6 W. L. Nicholson et al., "Resistance of *Bacillus* endospores to extreme terrestrial and extraterrestrial environments," *Microbiology and Molecular Biology Reviews* 64 (2000): 548–572.

7 E. Szathmáry and J. Maynard Smith, "The major evolutionary transitions," *Nature* 374 (1995): 227–232.

8 J. A. Lake, "Lynn Margulis (1938–2011)," *Nature* 480 (2011): 458.

9 L. Margulis, *Origin of Eukaryotic Cells* (New Haven: Yale University Press, 1971).

10 J. W. Costerton et al., "Microbial biofilms," *Annual Review of Microbiology* 49 (1995): 711–745.

11 For a lovely account of stromatolites in their natural habitat, go no further than Richard Fortey's book *Survivors: The Animals and Plants That Time Has Left Behind* (London: HarperCollins, 2011).

12 J. W. Schopf, "Fossil evidence of Archaean life," *Philosophical Transactions of the Royal Society B* 361 (2006): 869–885.

13 P. K. Singh et al., "Quorum-sensing signals indicate that cystic fibrosis lungs are infected with bacterial biofilms," *Nature* 407 (2000): 762–764.

14 M. W. Gray et al., "The origin and early evolution of mitochondria," *Genome Biology* 2 (2001), http://genomebiology.com/2001/2/6/reviews/1018.

15 G. I. McFadden and G. G. van Dooren, "Red algal genome affirms a common origin of all plastids," *Current Biology* 14 (2004): R514–R516.

16 T. Kleine et al., "DNA transfer from organelles to the nucleus: The idiosyncratic genetics of endosymbiosis," *Annual Review of Plant Biology* 60 (2009): 115–138.

17 G. I. McFadden and P. Gilson, "Something borrowed, something green: Lateral transfer of chloroplasts by secondary endosymbiosis," *Trends in Ecology and Evolution* 10 (1995): 12–17; G. I. McFadden, "Primary and secondary endosymbiosis and the origin of plastids," *Journal of Phycology* 37 (2001): 951–959.

18 F. Martin et al., "The genome of *Laccaria bicolor* provides insights into mycorrhizal symbiosis," *Nature* 452 (2008): 88–92.

19 J. Whitfield, "Fungal roles in soil ecology: Underground networking," *Nature* 449 (2007): 136–138.

20 Perhaps the most elegant way to lose your lunch is to read Carl Zimmer's book *Parasite Rex* (New York: Touchstone, 2001).

21 J. H. Day, "The life history of *Sacculina*," *Quarterly Journal of Microscopical Science* 77 (1935): 549–583.

22 S. T. Cole et al., "Deciphering the biology of *Mycobacterium tuberculosis* from the complete genome sequence," *Nature* 393 (1998): 537–544; S. T. Cole et al., "Massive gene decay in the leprosy bacillus," *Nature* 409 (2001): 1007–1011.

23 Carl Zimmer has followed his book on parasites (*Parasite Rex*) with one on viruses—*A Planet of Viruses* (Chicago: University of Chicago Press, 2012).

24 D. Raoult et al., "The 1.2-megabase genome sequence of mimivirus," *Science* 306 (2004): 1344–1350.

25 B. La Scola et al., "The virophage as a unique parasite of the giant mimivirus," *Nature* 455 (2008): 100–104.

26 The natural history of LINEs and SINEs can be found in E. S. Lander et al., "Initial sequencing and analysis of the human genome," *Nature* 409 (2001): 879–880.

27 R. O. Prum and A. H. Brush, "The evolutionary origin and diversification of feathers," *Quarterly Review of Biology* 77 (2002): 261–295.

28 P. M. O'Connor and L. P. A. M. Claessens, "Basic avian pulmonary design and flow-through ventilation in non-avian theropod dinosaurs," *Nature* 436 (2005): 253–256; F. E. Novas and P. F. Puertat, "New evidence concerning avian origins from the Late Cretaceous of Patagonia," *Nature* 387 (1997): 390–392; M. A. Norell et al., "A *Velociraptor* wishbone," *Nature* 389 (1997): 447; M. A. Norell et al., "A nesting dinosaur," *Nature* 378 (1995): 774–776.

29 For a review see M. A. Norell and X. Xu, "Feathered dinosaurs," *Annual Review of Earth and Planetary Sciences* 33 (2005): 277–299.

30 X. Xu et al., "The smallest known non-avian theropod dinosaur," *Nature* 408 (2000): 705–708; X. Xu et al., "A dromaeosaurid dinosaur with a filamentous integument from the Yixian Formation of China," *Nature* 401 (1999): 262–266.

31 X. Xu et al., "A gigantic bird-like dinosaur from the Late Cretaceous of China," *Nature* 447 (2007): 844–847.

32 When Gee Minor was a little older—about four—I took her to the Natural History Museum in London to see a traveling exhibition on the new and

unfamiliar feathered dinosaurs of China. I had published reports on most of the specimens on display, but had seen them only in photographs, so was excited to see them in real life, or, as it may be, death. One had to pay to get into this special exhibition, and it was tucked away in a side gallery—so it was away from the main dinosaur exhibit and thus patronized by very few visitors. The exhibit consisted of just nine specimens, moodily lit. As I became engrossed in the study of each fossil, Gee Minor whizzed around from fossil to fossil in the manner of a bumblebee flitting between flowers in a herbaceous border. A fossil that particularly engaged my attention was *Caudipteryx*, preserved on a large tabletop slab, a creature previously featured in *Nature* under my watch (J. Qiang et al., "Two feathered dinosaurs from northeastern China," *Nature* 393 [1998]: 753–761). As I was examining the specimen, deep in thought, a little face popped up on the other side of the slab and said, "Dad, did you punish this in *Nature*?" Out of the mouths of babes.

33 For an overview of *Archaeopteryx*, see P. Shipman, *Taking Wing: Archaeopteryx and the Origin of Bird Flight* (New York: Touchstone, 1999).

34 P. Domínguez Alonso et al., "The avian nature of the brain and inner ear of *Archaeopteryx*," *Nature* 430 (2004): 666–669.

35 P.-J. Chen et al., "An exceptionally well-preserved theropod dinosaur from the Yixian Formation of China," *Nature* 391 (1998): 147–152.

36 X. Xu et al., "An *Archaeopteryx*-like theropod from China and the origin of Avialae," *Nature* 475 (2011): 465–470; L. Witmer, "An icon knocked from its perch," *Nature* 475 (2011): 458–459.

37 F. Zhang et al., "A bizarre Jurassic maniraptoran from China with elongate ribbon-like feathers," *Nature* 455 (2008): 1105–1108.

38 C. Y. McLean et al., "Human-specific loss of regulatory DNA and the evolution of human-specific traits," *Nature* 471 (2011): 216–219.

39 N. Humphrey et al., "Human hand-walkers: Five siblings who never stood up" (2005), London School of Economics and Political Science Research Online, http://eprints.lse.ac.uk/id/eprint/463.

40 W. Enard et al., "Molecular evolution of *FOXP2*, a gene involved in speech and language," *Nature* 418 (2002): 869–872.

CHAPTER 4

1 G. Weber, "Top languages: The world's 10 most influential languages" (2008), http://www.andaman.org/BOOK/reprints/weber/rep-weber.htm, accessed 31 March 2012.

2 For an absorbing account of the history of the British Empire, see N. Ferguson, *Empire: The Rise and Demise of the British World Order and the Lessons for Global Power* (London: Allen Lane, 2002).

3 Niall Ferguson looked at America, too, in *Colossus: The Rise and Fall of the American Empire* (London: Penguin, 2004).

4 F. McLynn, *1759: The Year Britain Became Master of the World* (London: Vintage, 2008).

5 N. Ferguson, ed., *Virtual History: Alternatives and Counterfactuals* (London: Picador, 1997).

6 D. W. Meinig, *The Shaping of America: Atlantic America, 1492–1800* (New Haven: Yale University Press, 1986); D. W. Meinig, *The Shaping of America: Continental America, 1800–67* (New Haven: Yale University Press, 1993).

7 M. Alexander, *Old English Literature* (New York: Schocken, 1983).

8 I have two editions of the poem, separated in time by almost exactly a century. The first is a second edition (1898) of the standard text by A. J. Wyatt; the second is the text with the recent translation by the poet Seamus Heaney (London: Faber and Faber, 1999).

9 J. R. R. Tolkien, *Finn and Hengest: The Fragment and the Episode*, ed. Alan Bliss (London: HarperCollins, 1982).

10 S. McBrearty and N. Jablonski, "First fossil chimpanzee," *Nature* 437 (2005): 105–108.

11 Z. Zhou et al., "An exceptionally preserved Lower Cretaceous ecosystem," *Nature* 421 (2003): 807–814.

12 This was kindly demonstrated to me by Jean-Bernard Caron of the Royal Ontario Museum in Toronto.

13 D. Dashzeveg et al., "Extraordinary preservation in a new vertebrate assemblage from the Late Cretaceous of Mongolia," *Nature* 374 (1995): 446–449.

14 P. A. Allison and D. E. G. Briggs, "Exceptional fossil record: Distribution of soft-tissue preservation through the Phanerozoic," *Geology* 21 (1993): 527–530.

15 R. S. Sansom et al., "Non-random decay of chordate characters causes bias in fossil interpretation," *Nature* 463 (2010): 797–800.

16 G. Borgonie et al., "Nematoda from the terrestrial deep subsurface of South Africa," *Nature* 474 (2011): 79–82.

17 N. A. Cobb, *Nematodes and Their Relationships*, United States Department of Agriculture Yearbook (Washington, DC: US Department of Agriculture, 1914), 472. I am grateful to Roderic Page for tracing this quote.

18 G. Poinar Jr., "Trends in the evolution of insect parasitism by nematodes as inferred from fossil evidence," *Journal of Nematology* 35 (2003): 129–132.

19 G. Poinar Jr. and A. J. Boucot, "Evidence of intestinal parasites of dinosaurs," *Parasitology* 133 (2006): 245–249.

20 E. P. Hoberg et al., "Out of Africa: Origins of the *Taenia* tapeworms in humans," *Proceedings of the Royal Society of London B* 268 (2001): 781–787.

21 It is possible that *Opabinia* is obscurely related to velvet worms (onychophores) and tardigrades (water bears). G. E. Budd, "The morphology of *Opabinia regalis* and the reconstruction of the arthropod stem group," *Lethaia* 29 (1996): 1–14.

22 For a general review see *The Adequacy of the Fossil Record*, ed. S. K. Donovan and C. R. C. Paul (Chichester, UK: Wiley, 1998). The thesis of this book is

that the incompleteness of the fossil record need not dent its adequacy for solving a variety of paleontological problems.

23 The literature on the Doushantuo phosphorites is voluminous. For two recent articles see J. A. Cunningham et al., "Distinguishing geology from biology in the Ediacaran Doushantuo biota relaxes constraints on the timing of the origin of bilaterians," *Proceedings of the Royal Society of London B* 279 (2012): 2369–2376; and G. Jiang et al., "Stratigraphy and paleogeography of the Ediacaran Doushantuo Formation (ca. 635–551 Ma) in South China," *Gondwana Research* 19 (2011): 831–849.

24 F. H. T. Rhodes and R. Phillips, "The zoological affinities of conodonts," *Biological Reviews* 29 (1954): 419–452.

25 S. Conway Morris, "*Typhloesus wellsi* (Melton and Scott, 1973), a bizarre metazoan from the Carboniferous of Montana, U.S.A.," *Philosophical Transactions of the Royal Society of London B* 327 (1990): 595–624.

26 R. J. Aldridge et al., "The anatomy of conodonts," *Philosophical Transactions of the Royal Society of London B* 340 (1993): 405–421; R. J. Aldridge et al., "The affinities of conodonts: New evidence from the Carboniferous of Edinburgh, Scotland," *Lethaia* 19 (1986): 279–291.

27 A. Blieck et al., "Fossils, histology, and phylogeny: Why conodonts are not vertebrates," *Episodes* 33 (2010): 234–241; A. Kemp, "Hyaline tissue of thermally unaltered conodont elements and the enamel of vertebrates," *Alcheringa* 2 (2008): 23–36.

28 J. J. Sepkoski, "A compendium of fossil marine animal genera," *Bulletins of American Paleontology* 363 (2002): 1–560. It lives on as the Paleobiology Database, http://paleodb.org/cgi-bin/bridge.pl.

29 See for example J. Alroy et al., "Phanerozoic trends in the global diversity of marine invertebrates," *Science* 321 (1998): 97–100; J. Alroy, "The shifting balance of diversity among major marine animal groups," *Science* 329 (2010): 1191–1194; C. R. Marshall, "Marine biodiversity dynamics over deep time," *Science* 329 (2010): 1156–1157.

30 Paleontologists express this uncertainty as the "Signor-Lipps effect," after the two paleontologists who discussed the issues. See P. W. Signor III and J. H. Lipps, "Sampling bias, gradual extinction patterns, and catastrophes in the fossil record," in L. T. Silver and P. H. Schultz, eds., *Geological implications of impacts of large asteroids and comets on the earth*, Geological Society of America Special Publication 190 (Boulder, CO: Geological Society of America, 1982), 291–296.

31 See for example S. E. Peters, "Environmental determinants of extinction selectivity in the fossil record," *Nature* 454 (2008): 626–629; S. E. Peters and M. Foote, "Determinants of extinction in the fossil record," *Nature* 416 (2002): 420–424; A. B. Smith, "Large-scale heterogeneity of the fossil record: Implications for Phanerozoic biodiversity studies," *Philosophical Transactions of the Royal Society of London B* 356 (2001): 351–367; P. M. Barrett et al., "Dinosaur diversity and the rock record," *Proceedings of the Royal*

Society of London B 276 (2009): 2667–2674; G. T. Lloyd et al., "Quantifying the deep-sea rock and fossil record bias using coccolithophores," *Geological Society, London, Special Publications* 358 (2011): 167–177.

32 A few are known from rocks of the same age in nearby Orkney.

33 See for example W. J. Sollas and I. G. B. Sollas, "An account of the Devonian fish, *Palaeospondylus gunni*, Traquair," *Philosophical Transactions of the Royal Society of London B* 196 (1904): 267–294; J. A. Moy-Thomas, "The Devonian fish *Palaeospondylus gunni* Traquair," *Philosophical Transactions of the Royal Society of London B* 230 (1940): 391–413; K. S. Thomson et al., "A larval Devonian lungfish," *Nature* 426 (2003): 833–834; M. J. Newman and J. L. Den Blaauwen, "New information on the enigmatic Devonian vertebrate *Palaeospondylus gunni*," *Scottish Journal of Geology* 44 (2008): 89–91.

CHAPTER 5

1 R. Lee, "The outlook for population growth," *Science* 333 (2011): 569–573.

2 T. Benton, "Oceans of garbage," *Nature* 352 (1991): 113.

3 H. Kaessmann et al., "Great ape DNA sequences reveal a reduced diversity and an expansion in humans," *Nature Genetics* 27 (2001): 155–156; H. Kaessmann et al., "Extensive nuclear DNA sequence diversity among chimpanzees," *Science* 286 (1999): 1159–1162; Chimpanzee Sequencing and Analysis Consortium, "Initial sequence of the chimpanzee genome and comparison with the human genome," *Nature* 437 (2005): 69–87.

4 C. D. Huff et al., "Mobile elements reveal small population size in ancient ancestors of *Homo sapiens*," *Proceedings of the National Academy of Sciences of the USA* 107 (2010): 2147–2152. I should perhaps add a note here about the term "effective population size" used in studies such as this. The effective population size (N_e) is not the gross number of individuals in a population, but the number that contribute genes to the next generation. This is typically a much smaller number, given that many individuals in a population will die before reproducing, and a few individuals (such as dominant males) will contribute disproportionately to the gene pool—but it's the number that matters when one is tracking changes in genetic diversity.

5 S. R. Copeland et al., "Strontium isotope evidence for landscape use by early hominins," *Nature* 474 (2011): 76–78.

6 A. Manica et al., "The effect of ancient population bottlenecks on human phenotypic variation," *Nature* 448 (2007): 346–348.

7 M. H. Wolpoff, "Competitive exclusion among Lower Pleistocene hominids: The single species hypothesis," *Man* 6 (1971): 601–614.

8 R. E. F. Leakey and A. C. Walker, "*Australopithecus*, *Homo* and the single species hypothesis," *Nature* 261 (1976): 572–574.

9 K. Harvati et al., "The Later Stone Age Calvaria from Iwo Eleru, Nigeria: Morphology and chronology," *PLOS One* 6 (2011): e24024, doi:10.1371/journal.pone.0024024.

10 See for example R. Dennell, "Early *Homo sapiens* in China," *Nature* 468 (2010): 512–513; Y. Hou and L. X. Zhao, "An archeological view for the presence of early humans in China," *Quaternary International* 223–224 (2010): 10–19; X. Gao et al., "Revisiting the origin of modern humans in China and its implications for global human evolution," *Science China: Earth Sciences* 53 (2010): 1927–1940; and C. Shen et al., "The earliest hominin occupations of the Nihewan Basin of northern China: Recent progress in field investigations," *Asian Paleoanthropology* (2010): 169–180.

11 D. Reich et al., "Genetic history of an archaic hominin group from Denisova Cave in Siberia," *Nature* 468 (2010): 1053–1060.

12 M. Stoneking and J. Krause, "Learning about human population history from ancient and modern genomes," *Nature Reviews Genetics* 12 (2011): 603–614.

13 D. Curnoe, "A 150-year conundrum: Cranial robusticity and its bearing on the origin of Aboriginal Australians," *International Journal of Evolutionary Biology* (2011), doi:10.4061/2011/632484.

14 A. Urbain, "Le kou-prey ou bœuf gris cambodgien," *Bulletin de la Société Zoologique de France* 62 (1937): 305–307.

15 M. V. Erdmann et al., "Indonesian 'king of the sea' discovered," *Nature* 395 (1998): 335.

16 V. V. Dung et al., "A new species of living bovid from Vietnam," *Nature* 363 (1993): 443–445.

17 W. Robichaud, "Physical and behavioral description of a captive saola, *Pseudoryx nghetinhensis*," *Journal of Mammalogy* 79 (1998): 394–405.

18 I have sometimes wondered whether I might compile a list of antelopes and other ungulates (the okapi isn't an antelope, but a relative of the giraffe) whose names would be useful in such a context. I once placed ADDAX in a competitive setting, much to the chagrin of my opponent, who challenged it and lost. Others worth keeping up your sleeve are GNU, TOPI, KOB, ELAND, PUDU, KUDU, BOK, NYALA, NILGAI, IMPALA, and ORYX. I am sure you can think of lots more.

19 "Notes," *Nature* 64 (1901): 188.

20 See for example J. B. Buhs, *Bigfoot: The Life and Times of a Legend* (Chicago: University of Chicago Press, 2009).

21 There are many recent, excellent accounts of human evolution, and the material in this chapter draws on several. I recommend (in no particular order) C. Stringer, *The Origin of Our Species* (London: Allen Lane, 2011); D. Falk, *The Fossil Chronicles: How Two Controversial Discoveries Changed Our View of Human Evolution* (Berkeley: University of California Press, 2011); J. Reader, *Missing Links: In Search of Human Origins* (Oxford: Oxford University Press, 2011); and A. Gibbons, *The First Human: The Race to Discover Our Earliest Ancestors* (New York: Anchor, 2007). For the student there is the 2011–2012 edition of R. Jurmain et al., *Introduction to Physical Anthropology* (Belmont, CA: Wadsworth Cengage Learning)—but for your coffee table it

would be hard to beat A. Roberts, *Evolution: The Human Story* (London: Dorling Kindersley, 2011).

22 P. Shipman, *The Man Who Found the Missing Link: The Extraordinary Life of Eugène Dubois* (New York: Simon and Schuster, 2001).

23 A. C. Haddon, "*Eoanthropus dawsoni*," *Science* 37 (1913): 91–92.

24 I have borrowed the term "Piltdown committee" and drawn on Dean Falk's excellent account of events in *The Fossil Chronicles: How Two Controversial Discoveries Changed Our View of Human Evolution* (Berkeley: University of California Press, 2011). Falk's account of the possible suppression of Dart's 1929 monograph on *Australopithecus africanus*—which remains unpublished to this day—is especially noteworthy.

25 Those more used to baseball are invited to substitute the appropriate metaphors.

26 R. Dart, "*Australopithecus africanus*: The man-ape of South Africa," *Nature* 115 (1925): 195–199.

27 See the correspondence column in *Nature* 115 (1925): 234–236 for immediate reactions from several members of the Piltdown committee, notably Sir Arthur Keith and Arthur Smith-Woodward, who had to some extent staked their reputations on Piltdown. The debate between Dart and Keith became somewhat acrimonious, as shown by the exchange in *Nature* 116 (1925): 462–463.

28 R. Broom, "Some notes on the Taungs skull," *Nature* 115 (1925): 569–571.

29 R. Broom, "A new fossil anthropoid skull from South Africa," *Nature* 138 (1936) 486–488; R. Broom, "The dentition of *Australopithecus*," *Nature* 138 (1936): 719; R. Broom, "Discovery of a lower molar of *Australopithecus*," *Nature* 140 (1937): 681–682; R. Broom, "Discovery of teeth of *Australopithecus*," *Nature* 140 (1937): 680; R. Broom, "More discoveries of *Australopithecus*," *Nature* 141 (1938): 828–829; R. Broom, "The Pleistocene anthropoid apes of South Africa," *Nature* 142 (1938): 377–379; R. Broom, "Further evidence on the structure of the South African Pleistocene anthropoids," *Nature* 142 (1938): 897–899; R. Broom, "Structure of the Sterkfontein ape," *Nature* 147 (1941): 86; R. Broom, "Mandible of a young *Paranthropus* child," *Nature* 147 (1941): 607–608; R. Broom, "The origin of man," *Nature* 148 (1941): 10–14; R. Broom, "The hand of the ape-man, *Paranthropus robustus*," *Nature* 149 (1942): 513–514; R. Broom, "An ankle-bone of the ape-man, *Paranthropus robustus*," *Nature* 152 (1945): 389–390; R. Broom, "The upper milk molars of the ape-man, *Plesianthropus*," *Nature* 159 (1947): 602; R. Broom and J. T. Robinson, "Size of the brain in the ape-man, *Plesianthropus*," *Nature* 161 (1948): 438; R. Broom, "Another new type of fossil ape-man," *Nature* 163 (1949): 57; and R. Broom and J. T. Robinson, "Eruption of the permanent teeth in the South African fossil ape-men," *Nature* 167 (1951): 443. Broom died, vindicated but still publishing, at the age of eighty-four—his obituary by W. E. LeGros Clark appeared in *Nature* 167 (1951): 752. Broom's discoveries were, like Dart's, not immune to the view

that they were more likely to be fossil apes than hominins—see the letter by E. Schwarz, *Nature* 138 (1936): 969, countered by Broom, *Nature* 139 (1937): 326.

30 Even Sir Arthur Keith was won over, with a gracious admission in *Nature* 159 (1947): 277.

31 D. Black, "Tertiary man in Asia—The Chou Kou Tien discovery," *Bulletin of the Geological Society of China* 5 (1926): 207–208.

32 D. Black, *On an Adolescent Skull of* Sinanthropus pekinensis *in Comparison with an Adult of the Same Species and with Other Hominid Skulls Recent and Fossil* (Peking: Geological Survey of China, 1931); F. Weidenreich, "The new discovery of three skulls of *Sinanthropus pekinensis*," *Nature* 139 (1937): 269–272.

33 P. Teilhard de Chardin and W. G. Pei, "The lithic industry of the *Sinanthropus* deposits in Choukoutien," *Bulletin of the Geological Society of China* 11 (1932): 315–364; D. Black, "Evidences of the use of fire by *Sinanthropus*," *Bulletin of the Geological Society of China* 11 (1932): 107–108.

34 G. H. R. von Koenigswald and F. Weidenreich, "The relationship between *Pithecanthropus* and *Sinanthropus*," *Nature* 144 (1939): 926–929.

35 J. S. Weiner et al., "The solution of the Piltdown problem," *Bulletin of the British Museum (Natural History) Geology* 2 (1953): 139–146.

36 B. G. Gardiner, "The Piltdown forgery: A re-statement of the case against Hinton," *Zoological Journal of the Linnean Society* 139 (2003): 315–335.

37 V. Morell, *Ancestral Passions: The Leakey Family and the Quest for Humankind's Beginnings* (New York: Touchstone, 1996).

38 L. S. B. Leakey, "A new fossil skull from Olduvai," *Nature* 184 (1959): 491–493.

39 L. S. B. Leakey et al., "A new species of the genus *Homo* from Olduvai Gorge," *Nature* 202 (1964): 7–9.

40 R. E. F. Leakey, "Evidence for an advanced Plio-Pleistocene hominid from East Rudolf, Kenya," *Nature* 242 (1973): 447–450.

41 D. Curnoe, "A review of early *Homo* in southern Africa focusing on cranial, mandibular and dental remains, with the description of a new species (*Homo gautengensis*, sp. nov.)," *Homo: Journal of Comparative Human Biology* 61 (2010): 151–177.

42 Some recently discovered fossils confirm that there were at least two kinds of early *Homo*. See M. G. Leakey et al., "New fossils from Koobi Fora in northern Kenya confirm taxonomic diversity in early *Homo*," *Nature* 488 (2012): 201–204.

43 B. Wood and M. Collard, "The human genus," *Science* 284 (1999): 65–71.

44 L. R. Berger et al., "*Australopithecus sediba*: A new species of *Homo*-like australopith from South Africa," *Science* 328 (2010): 195–204.

45 S. Semaw et al., "2.5-million-year-old stone tools from Gona, Ethiopia," *Nature* 385 (1997): 333–336; S. P. McPherron et al., "Evidence for stone-tool-assisted consumption of animal tissues before 3.39 million years ago at Dikika, Ethiopia," *Nature* 466 (2010): 857–860.

46 J.-L. Arsuaga, "Three new human skulls from the Sima de los Huesos Middle Pleistocene site in Sierra de Atapuerca, Spain," *Nature* 362 (1993): 534–537.

47 J. M. Bermúdez de Castro, "A hominid from the Lower Pleistocene of Atapuerca, Spain: Possible ancestor to Neandertals and modern humans," *Science* 276 (1997): 1392–1395. For a general review of hominin variation and taxonomy during this period of time, see G. P. Rightmire, "*Homo* in the Middle Pleistocene: Hypodigms, variation and species recognition," *Evolutionary Anthropology* 17 (2008): 8–21.

48 F. Brown et al., "Early *Homo erectus* skeleton from west Lake Turkana, Kenya," *Nature* 316 (1985): 788–792.

49 Once again, for a discussion of these issues see B. Wood and M. Collard, "The human genus," *Science* 284 (1999): 65–71.

50 C. J. Lepre et al., "An earlier origin for the Acheulian," *Nature* 477 (2011): 82–85.

51 R. Ferring et al., "Earliest human occupation of Dmanisi (Georgian Caucasus) dated to 1.85–1.78 Ma," *Proceedings of the National Academy of Sciences of the USA* 108 (2011): 10432–10436.

52 D. Lordkipanidze et al., "Postcranial evidence from early *Homo* from Dmanisi, Georgia," *Nature* 449 (2007): 305–310; L. Gabunia et al., "Découverte d'un nouvel hominidé à Dmanissi (Transcaucasie, Géorgie)," *Comptes Rendus Palevol* 1 (2002): 243–254.

53 C. J. Lepre et al., "An earlier origin for the Acheulian," *Nature* 477 (2011): 82–85.

54 R. Dennell and W. Roebroeks, "An Asian perspective on early human dispersal from Africa," *Nature* 438 (2005): 1099–1104.

55 I. McDougall et al., "Stratigraphic placement and age of modern humans from Kibish, Ethiopia," *Nature* 433 (2005): 733–736.

56 C. W. Marean et al., "Early human use of marine resources and pigment in South Africa during the Middle Pleistocene," *Nature* 449 (2007): 905–908.

57 R. L. Cann et al., "Mitochondrial DNA and human evolution," *Nature* 325 (1987): 31–36.

58 J. Wainscoat, "Out of the Garden of Eden," *Nature* 325 (1987): 13.

59 R. E. Green et al., "A draft sequence of the Neandertal genome," *Science* 328 (2010): 710–722.

60 D. Reich et al., "Genetic history of an archaic hominin group from Denisova Cave in Siberia," *Nature* 468 (2010): 1053–1060.

61 M. F. Hammer et al., "Genetic evidence for archaic admixture in Africa," *Proceedings of the National Academy of Science of the USA* 108 (2011): 15123–15128; K. Harvati et al., "The Later Stone Age Calvaria from Iwo Eleru, Nigeria: Morphology and chronology," *PLOS One* 6 (2011): e24024, doi:10.1371/journal.pone.0024024.

62 I've always been puzzled by Genesis 4:17: "And Cain knew his wife; and she conceived, and bare Enoch: and he builded a city, and called the name of

the city, after the name of his son, Enoch." Where did Cain find his wife, and all the other citizens of Enoch? Obviously, they were already there, yet unrecorded, outside Cain's African Eden.

63 C. Howell, "Omo research expedition," *Nature* 219 (1968): 567–572.

64 D. C. Johanson and M. Taieb, "Plio-Pleistocene hominid discoveries in Hadar, Ethiopia," *Nature* 260 (1976): 293–297.

65 T. D. White et al., "*Australopithecus ramidus*, a new species of hominid from Aramis, Ethiopia," *Nature* 371 (1994): 306–312.

66 T. D. White et al., "*Ardipithecus ramidus* and the paleobiology of early hominids," *Science* 326 (2009): 75–86.

67 M. G. Leakey et al., "New four-million-year-old hominid species from Kanapoi and Allia Bay, Kenya," *Nature* 376 (1995): 565–571; Y. Haile-Selassie, "Late Miocene hominids from the Middle Awash, Ethiopia," *Nature* 412 (2001): 178–181; B. Senut et al., "First hominid from the Miocene (Lukeino Formation, Kenya)," *Comptes rendus de l'Académie des sciences*, series 2a, 332 (2001): 137–144.

68 B. Wood and T. Harrison, "The evolutionary context of the first hominins," *Nature* 470 (2011): 347–352.

69 M. Brunet et al., "A new hominid from the Upper Miocene of Chad, Central Africa," *Nature* 418 (2002): 145–151.

70 C. P. E. Zollikofer et al., "Virtual cranial reconstruction of *Sahelanthropus tchadensis*," *Nature* 434 (2005): 755–759.

71 M. H. Wolpoff et al., "*Sahelanthropus* or '*Sahelpithecus*'?," *Nature* 419 (2002): 581–582; M. Brunet et al., "*Sahelanthropus* or '*Sahelpithecus*'?," *Nature* 419 (2002): 582.

CHAPTER 6

1 Equotes, http://bevets.com/equotesg.htm#G, accessed 4 April 2012.

2 S. Reuland, *Sunbeams from Cucumbers*, http://stevereuland.blogspot.com/2006/04/wittlessly-quote-mining.html, accessed 4 April 2012.

3 Psalms 14:1. I am grateful to Andrew Thaler for pointing out that particular gem.

4 J. Conard, "Palaeolithic ivory sculptures from southwestern Germany and the origins of figurative art," *Nature* 426 (2003): 830–832.

5 Although the recent news story about a Jewish religious court that sentenced a dog to death by stoning was, apparently, a hoax. http://newsfeed.time.com/2011/06/18/shocking-sentence-jewish-court-condemns-dog-to-death-by-stoning/, accessed 6 November 2012.

6 I first found this story in Felipe Fernández-Armesto, *So You Think You're Human? A Brief History of Humankind* (Oxford: Oxford University Press, 2004).

CHAPTER 7

1 M. Srinivasan and A. Ruina, "Computer optimization of a minimal biped model discovers walking and running," *Nature* 439 (2006): 72–75.

2 W. E. H. Harcourt-Smith and L. C. Aiello, "Fossils, feet and the evolution of human bipedal locomotion," *Journal of Anatomy* 204 (2004): 403–416; D. L. Gebo, "Climbing, brachiation, and terrestrial quadrupedalism: Historical precursors of hominid bipedalism," *American Journal of Physical Anthropology* 101 (1996): 55–92; C. O. Lovejoy, "Evolution of human walking," *Scientific American*, November 1988, 118–125.

3 K. D. Hunt, "The evolution of human bipedality: Ecology and functional morphology," *Journal of Human Evolution* 26 (1994): 183–202.

4 P. E. Wheeler, "The thermoregulatory advantages of hominid bipedalism in open equatorial environments: The contribution of increased convective heat loss and cutaneous evaporative cooling," *Journal of Human Evolution* 21 (1991): 107–115; P. E. Wheeler, "The evolution of bipedality and loss of functional body hair in hominids," *Journal of Human Evolution* 13 (1984): 91–98.

5 E. Morgan, *The Aquatic Ape Hypothesis* (London: Souvenir Press, 1997).

6 Neil H. Shubin of the University of Chicago told me a story about a drive in Morocco, during which he and the other passengers noticed a tree in the distance that had been colonized by what looked like large birds. Vultures, maybe? As they got closer, it became clear that the birds were in fact a herd of goats that filled the entire tree from the trunk to the outermost twigs.

7 J. Diamond, *The Rise and Fall of the Third Chimpanzee* (London: Random House, 1991); J. Diamond, *Why Is Sex Fun? The Evolution of Human Sexuality* (London: Weidenfeld and Nicolson, 1997).

8 M. Andersson and Y. Iwasa, "Sexual selection," *Trends in Ecology and Evolution* 11 (1996): 53–58; J. Maynard Smith, "Theories of sexual selection," *Trends in Ecology and Evolution* 6 (1991): 146–151.

9 W. D. Hamilton and M. Zuk, "Heritable true fitness and bright birds: A role for parasites?," *Science* 218 (1982): 384–387.

10 R. A. Fisher, *The Genetical Theory of Natural Selection* (Oxford: Clarendon Press, 1930).

11 A. Zahavi, "Mate selection: A selection for a handicap," *Journal of Theoretical Biology* 53 (1975): 205–214.

12 At this point you will ask yourself why expensive sports cars are seen as sexy, whereas more inexpensive and practical station wagons are not. *Or are they?* Advertisers, as we have seen, are very good instinctive judges of human motivation—so how would a copywriter sell the idea of station wagons to men of a certain age and maturity who still hanker after that red sports car? One answer—I forget the particular campaign—went like this. A man was pictured alongside his beautiful girlfriend in a snappy red roadster. In the next panel, the same man was pictured next to the same beautiful woman, now his adoring wife, in a station wagon full of children

and other appurtenances of middle-aged success—dogs, sports equipment, and so on. The caption went something like this: Sports cars are full of the empty show of youth. But now you've got the girl, fathered all these children, and achieved some status, you have obviously proved your virility. You will therefore need a suitable vehicle in which to display the fruits of your loins and your material acquisitions to males in sports cars, who have yet to achieve alpha-male dominance status.

13 The following paper links sex with bipedality, though the thesis is much more complicated than mine: S. T. Parker, "A sexual selection model for hominid evolution," *Human Evolution* 2 (1987): 235–253.

14 F. Szalay and R. K. Costello, "Evolution of permanent estrus displays in hominids," *Journal of Human Evolution* 20 (1991): 439–464.

15 L. Benshoof and R. Thornhill, "The evolution of monogamy and concealed ovulation in humans," *Journal of Social and Biological Structures* 2 (1979): 95–106.

16 H. Greiling and D. M. Buss, "Women's sexual strategies: The hidden dimension of extra-pair mating," *Personality and Individual Differences* 28 (2000): 929–963. In his book *The Rise and Fall of the Third Chimpanzee*, Jared Diamond relates an anecdote about the prevalence of extra-pair paternity related by an obstetrician. Basically, there's very much more of it than people either claim or are prepared to admit, given that the social norm is the maintenance of overt monogamy.

17 L. Bellamy and A. Pomiankowski, "Why promiscuity pays," *Nature* 479 (2011): 184–186; C. K. Cornwallis et al., "Promiscuity and the evolutionary transition to complex societies," *Nature* 466 (2010): 969–972; D. F. Westneat and I. R. Stewart, "Extra-pair paternity in birds: Causes, correlates, and conflict," *Annual Review of Ecology, Evolution and Systematics* 34 (2003): 365–396; S. C. Griffith et al., "Extra-pair paternity in birds: A review of interspecific variation and adaptive function," *Molecular Ecology* 11 (2002): 2195–2212; M. Petrie and B. Kempenaers, "Extra-pair paternity in birds: Explaining variation between species and populations," *Trends in Ecology and Evolution* 13 (1998): 52–58.

18 D. R. Rubenstein and I. J. Lovette, "Reproductive skew and selection on female ornamentation in social species," *Nature* 462 (2009): 786–789.

19 *Nature*, along with most other scientific journals these days, receives submissions online. It was not always so. I well remember the days when manuscripts arrived in the mail, and in quadruplicate, together with any supporting information, so that the office would have a copy, and there'd be one each for up to three potential referees. One day I received a huge parcel, containing a somewhat way-out manuscript on the origin of human secondary sexual characteristics. The author's thesis was supported by a particularly lurid example of what used to be called "men's magazines." There were four copies of this, too. I swear that I sent all of them back to the author. If it's in an archive somewhere, I'm not aware of it. Honest.

20 A. S. Jackson et al., "The effect of sex, age and race on estimating percentage body fat from body mass index: The Heritage Family Study," *International Journal of Obesity* 26 (2002): 789–796.

21 P. Frost, *Fair Women, Dark Men: The Forgotten Roots of Racial Prejudice* (Christchurch, New Zealand: Cybereditions, 2005).

22 D. W. Yu and G. H. Shepard, "Is beauty in the eye of the beholder?," *Nature* 396 (1998): 321–322.

23 D. R. Rubenstein and I. J. Lovette, "Reproductive skew and selection on female ornamentation in social species," *Nature* 462 (2009): 786–789.

24 F. D. Wyatt, "Fifty years of pheromones," *Nature* 457 (2009): 262–263.

25 Note that I didn't write "nobody." We humans can get turned on by the most peculiar things. For example, a friend of a friend reportedly sells rubber Wellington boots online to an eager market of fetishists; and it remains unknown, at least to me, why so many women of my acquaintance are so enraptured by, of all things, shoes.

26 See for example S. Dagenais et al., "A systematic review of low back pain cost of illness studies in the United States and internationally," *Spine Journal* 8 (2008): 8–20.

27 K. K. Whitcome et al., "Fetal load and evolution of lumbar lordosis in bipedal hominins," *Nature* 450 (2007): 1075–1078.

28 For a review see D. Lieberman, *The Evolution of the Human Head* (Cambridge, MA: Harvard University Press, 2011).

29 C. O. Lovejoy et al., "The pelvis and femur of *Ardipithecus ramidus*: The emergence of upright walking," *Science* 326 (2009): 71, 71e1–71e6.

30 C. V. Ward, "Interpreting the posture and locomotion of *Australopithecus afarensis*: Where do we stand?," in "Yearbook of Physical Anthropology," supplement, *American Journal of Physical Anthropology* 119, suppl. 35 (2002): 185–215.

31 D. M. Bramble and D. E. Lieberman, "Endurance running and the evolution of *Homo*," *Nature* 432 (2004): 345–352.

32 L. Rook et al., "*Oreopithecus* was a bipedal ape after all: Evidence from the iliac cancellous architecture," *Proceedings of the National Academy of Sciences of the USA* 96 (1999): 8795–8799; S. Moyà-Solà et al., "Evidence of hominid-like precision grip capability in the hand of the Miocene ape *Oreopithecus*," *Proceedings of the National Academy of Sciences of the USA* 96 (1999): 313–317.

33 T. Harrison, "A reassessment of the phylogenetic relationships of *Oreopithecus bambolii* Gervais," *Journal of Human Evolution* 15 (1986): 541–583.

CHAPTER 8

1 To use the argument on the illusion of complexity I presented in chapter 3, modern technology proceeds by the combination of parts that individually have become very simple. Yes, you probably could make something that

had the processing power of an iPad from vacuum tubes—but it would be enormous, incredibly expensive, shockingly unreliable, and wildly inefficient.

2 R. Wrangham, *Catching Fire: How Cooking Made Us Human* (London: Profile Books, 2009); L. C. Aiello and P. Wheeler, "The expensive-tissue hypothesis: The brain and the digestive system in human and primate evolution," *Current Anthropology* 36 (1995): 199–221.

3 A well-known phenomenon is the social facilitation of eating, in which people tend to eat more when in company than when dining alone. See for example V. I. Clendenen et al., "Social facilitation of eating among friends and strangers," *Appetite* 23 (1994): 1–13. Many years ago I wrote an account of this paper, or one very like it, as an excuse to tell a favorite joke. "Have another bagel, rabbi," says the hostess. "I couldn't possibly," says the rabbi, "I've already had three." "You've had four," the hostess replies, "but who's counting?"

4 T. Taylor, *The Artificial Ape: How Technology Changed the Course of Human Evolution* (New York: Palgrave Macmillan, 2010).

5 W. B. Arthur, *The Nature of Technology: What It Is and How It Evolves* (London: Allen Lane, 2009).

6 J. Bradshaw, *In Defence of Dogs: Why Dogs Need Our Understanding* (London: Allen Lane, 2011); P. Shipman, *The Animal Connection: A New Perspective on What Makes Us Human* (New York: W. W. Norton, 2011).

7 J. Diamond, "The double puzzle of diabetes," *Nature* 423 (2003): 599–602.

8 S. Semaw et al., "2.5-million-year-old stone tools from Gona, Ethiopia," *Nature* 385 (1997): 333–336; S. P. McPherron et al., "Evidence for stone-tool-assisted consumption of animal tissues before 3.39 million years ago at Dikika, Ethiopia," *Nature* 466 (2010): 857–860.

9 See for example G. R. Hunt, "Manufacture and use of hook-tools by New Caledonian crows," *Nature* 379 (1996): 249–251.

10 A. Whiten, "The second inheritance system of chimpanzees and humans," *Nature* 437 (2005): 52–55.

11 M. Haslam et al., "Primate archaeology," *Nature* 460 (2009): 339–344.

12 This discussion throws a whole new light on the words to that otherwise utterly infuriating song "There's a Hole in My Bucket, Dear Henry."

13 See for example A. H. Taylor et al., "Do New Caledonian crows solve physical problems through causal reasoning?," *Proceedings of the Royal Society of London B* 276 (2009): 247–254; A. H. Taylor et al., "Complex cognition and behavioural innovation in New Caledonian crows," *Proceedings of the Royal Society of London B* 277 (2010): 2637–2643; A. H. Taylor et al., "Context-dependent tool use in New Caledonian crows," *Biology Letters* (2011), doi:10.1098/rsbi.2011.0782.

14 The New Caledonian crow, for example, has a large brain relative to its size, even compared with other crow species. See J. Cnotka et al., "Extraordinary

large brains in tool-using New Caledonian crows (*Corvus moneduloides*)," *Neuroscience Letters* 433 (2008): 241–245; J. Mehlhorn et al., "Tool-making New Caledonian crows have large associative brain areas," *Brain, Behaviour and Evolution* 75 (2010): 63–70.

15 C. J. Lepre et al., "An earlier origin for the Acheulian," *Nature* 477 (2011): 82–85.

16 G. Sharon, "Acheulian giant-core technology: A worldwide perspective," *Current Anthropology* 50 (2009): 335–367; S. J. Lycett et al., "Acheulean variability and hominin dispersals: A model-bound approach," *Journal of Archaeological Science* 35 (2008): 553–562.

17 T. Wynn, "Handaxe enigmas," *World Archaeology* 27 (1995): 10–24; J. C. Whittaker and G. McCall, "Handaxe-hurling hominids: An unlikely story," *Current Anthropology* 42 (2001): 566–572.

18 S. Mithen, "Handaxes: The first aesthetic artefacts," in E. Voland and K. Grammer, eds., *Evolutionary Aesthetics* (Berlin: Springer-Verlag, 2003), 261–274; A. J. Machin et al., "Why are some handaxes symmetrical? Testing the influence of handaxe morphology on butchery effectiveness," *Journal of Archaeological Science* 34 (2007): 883–893.

19 A. Kohn and S. Mithen, "Handaxes: Products of sexual selection?," *Antiquity* 73 (1999): 518–526. See also the counterargument—A. Nowell and M. L. Chang, "The case against sexual selection as an explanation for handaxe morphology," *PaleoAnthropology* (2009): 77–88.

20 F. Berna et al., "Microstratigraphic evidence of in situ fire in the Acheulean strata of Wonderwerk Cave, Northern Cape province, South Africa," *Proceedings of the National Academy of Sciences of the USA* 109 (2012): 7593–7594.

21 A. Brumm et al., "Hominins on Flores, Indonesia, by one million years ago," *Nature* 464 (2010): 748–752.

22 C. Dean et al., "Growth processes in teeth distinguish modern humans from *Homo erectus* and earlier hominins," *Nature* 414 (2001): 628–631; H. Coqueugniot et al., "Early brain growth in *Homo erectus* and implications for cognitive ability," *Nature* 431 (2004): 299–302.

23 P. Mellars, "Neanderthals and the modern human colonization of Europe," *Nature* 432 (2004): 461–465.

24 B. Wood, "Origin and evolution of the genus *Homo*," *Nature* 355 (1992): 783–790; H. M. McHenry and K. Coffing, "*Australopithecus* to *Homo*: Transformations in body and mind," *Annual Review of Anthropology* 29 (2000): 125–146.

25 B. Wood and M. Collard, "The human genus," *Science* 284 (1999): 65–71.

26 B. Asfaw et al., "*Australopithecus garhi*: A new species of early hominid from Ethiopia," *Science* 284 (1999): 629–635; L. Berger et al., "*Australopithecus sediba*: A new species of *Homo*-like australopith from South Africa," *Science* 328 (2010): 195–204.

CHAPTER 9

1 University of Cambridge, Research News, http://www.cam.ac.uk/research/news/the-bird-tango-cambridge-academic-fuses-love-of-birds-and-dance/, accessed 12 April 2012.

2 For example, this clip narrated by David Attenborough, from BBC Worldwide, http://www.youtube.com/watch?v=BGPGknpq3eO, accessed 12 April 2012.

3 J. M. Dally et al., "Food-caching western scrub-jays keep track of who was watching when," *Science* 312 (2006): 1662–1665; C. R. Raby et al., "Planning for the future by western scrub-jays," *Nature* 445 (2007): 919–992; N. J. Emery and N. S. Clayton, "Effects of experience and social context on prospective caching strategies by scrub jays," *Nature* 414 (2001): 443–446.

4 N. J. Emery and N. S. Clayton, "Comparing the complex cognition of birds and primates," in L. J. Rogers and G. Kaplan, eds., *Comparative Vertebrate Cognition: Are Primates Superior to Non-Primates?* (New York: Kluwer Academic/Plenum, 2004), 3–56.

5 N. J. Emery, "Cognitive ornithology: The evolution of avian intelligence," *Philosophical Transactions of the Royal Society B* 361 (2006): 23–43; S. Shultz and R. I. M. Dunbar, "Social bonds in birds are associated with brain size and contingent on the correlated evolution of life-history and increased parental investment," *Biological Journal of the Linnean Society* 100 (2010): 111–123; M. J. Burish et al., "Brain architecture and social complexity in modern and ancient birds," *Brain, Behavior and Evolution* 63 (2004): 107–124.

6 H. J. Jerison, "The theory of encephalization," *Annals of the New York Academy of Sciences* 299 (1977): 146–160; H. J. Jerison and H. B. Barlow, "Animal intelligence as encephalization," *Philosophical Transactions of the Royal Society of London B* 308 (1985): 21–35.

7 With the possible exception, I like to think, of the gluteus maximus—the major muscle that forms the mass of the buttocks, essential to our bipedal stance. You may draw whatever conclusion you will from this comparison.

8 G. P. Rightmire, "Brain size and encephalization in early to mid-Pleistocene *Homo*," *American Journal of Physical Anthropology* 124 (2004): 109–123.

9 C. B. Ruff et al., "Body mass and encephalization in Pleistocene *Homo*," *Nature* 387 (1997): 173–176.

10 G. P. Rightmire, "Human evolution in the Middle Pleistocene: The role of *Homo heidelbergensis*," *Evolutionary Anthropology* 6 (1998): 218–227.

11 H. Thieme, "Lower Palaeolithic hunting spears from Germany," *Nature* 385 (1997): 807–810.

12 R. N. Carmody and R. W. Wrangham, "The energetic significance of cooking," *Journal of Human Evolution* 57 (2009): 379–391.

13 L. C. Aiello and P. Wheeler, "The expensive-tissue hypothesis: The brain and the digestive system in human and primate evolution," *Current Anthropology* 36 (1995): 199–221.

14 A. Navarrete et al., "Energetics and the evolution of human brain size," *Nature* 480 (2011): 91–93.

15 R. Wrangham, *Catching Fire: How Cooking Made Us Human* (London: Profile Books, 2009).

16 H. H. Stedman et al., "Myosin gene mutation correlates with anatomical changes in the human lineage," *Nature* 428 (2004): 415–418.

17 D. Lieberman, *The Evolution of the Human Head* (Cambridge, MA: Harvard University Press, 2011).

18 K. Hawkes et al., "Grandmothering, menopause, and the evolution of human life histories," *Proceedings of the National Academy of Sciences of the USA* 95 (1998): 1336–1339.

19 P. S. Kim et al., "Increased longevity evolves from grandmothering," *Proceedings of the Royal Society of London B* 279 (2012): 4880–4884.

20 R. Mace and A. Alvergne, "Female reproductive competition within families in rural Gambia," *Proceedings of the Royal Society of London B* 279 (2012): 2219–2227.

21 C. Stringer and C. Gamble, *In Search of the Neanderthals: Solving the Puzzle of Human Origins* (London: Thames and Hudson, 1994).

22 In *The Social Conquest of Earth*, Edward O. Wilson stresses the importance of the evolution of advanced social life, achieved in two very different ways, by *Homo sapiens* and by social insects such as ants and termites.

23 C. Spearman, "'General intelligence,' objectively determined and measured," *American Journal of Psychology* 15 (1904): 201–292; J. Duncan et al., "A neural basis for general intelligence," *Science* 289 (2000): 457–460; I. J. Deary et al., "Genetic contributions to stability and change in intelligence from childhood to old age," *Nature* 482 (2012): 212–215.

CHAPTER 10

1 That is, bad luck: one can't help but think that Hamlet would have had much sympathy with William Bell, who wrote the lyrics of the blues standard "Born under a Bad Sign": "If it wasn't for bad luck, I wouldn't have no luck at all."

2 I am grateful to Walter Gratzer for alerting me to this, along with many others in a similar vein, such as "Scientists Make Gorillas Pregnant."

3 M. J. Noad et al., "Cultural revolution in whale songs," *Nature* 408 (2000): 537.

4 M. S. Brainard and A. J. Doupe, "What songbirds teach us about learning," *Nature* 417 (2002): 351–358.

5 Now I am older and past my prime, I can play practically anything, even Whitesnake.

6 As reported by Humphrey Carpenter in *J. R. R. Tolkien: A Biography* (London: HarperCollins, 1977).

7 C. Henshilwood et al., "Emergence of modern human behavior: Middle Stone Age engravings from South Africa," *Science* 295 (2002): 1278–1280.

CHAPTER 11

1 When anyone uses the word "surely" in an argument, it usually means that they've had to resort to special pleading.

2 J. M. Dally et al., "Food-caching western scrub-jays keep track of who was watching when," *Science* 312 (2006): 1662–1665.

3 N. J. Emery and N. S. Clayton, "Effects of experience and social context on prospective caching strategies by scrub jays," *Nature* 414 (2001): 443–446.

4 S. Baron-Cohen et al., "Does the autistic child have a 'theory of mind'?," *Cognition* 21 (1985): 37–46.

5 S. Ramsden et al., "Verbal and non-verbal intelligence changes in the teenage brain," *Nature* 479 (2011): 113–116; K. Powell, "How does the teenage brain work?," *Nature* 442 (2006): 865–867.

6 In the notorious song by Harry "The Hipster" Gibson. I have a truly marvelous anecdote about this song, but this footnote is too small to contain it. And, in case you were wondering, the perpetrator wasn't Mr. Murphy. He was just as puzzled by the occurrence of nembutal in his overalls.

7 You can find it in English in Borges's collection, *Labyrinths*.

8 K. Smith, "Neuroscience vs philosophy: Taking aim at free will," *Nature* 477 (2011): 23–25.

9 See for example the interview with vision researcher Christof Koch in *Science* 335 (2012): 1426–1427.

10 N. J. Dominy and P. W. Lucas, "Ecological importance of trichromatic vision to primates," *Nature* 410 (2001): 363–366.

11 N. J. Dingemanse et al., "Behavioural reaction norms: Animal personality meets individual plasticity," *Trends in Ecology and Evolution* 25 (2010): 81–89; J. Stamps and T. G. G. Groothuis, "The development of animal personality: Relevance, concepts and perspectives," *Biological Reviews* 85 (2010): 301–325.

12 D. Cyranoski, "Pet project," *Nature* 466 (2010): 1036–1038.

13 M. Wolf et al., "Life-history trade-offs favour the evolution of animal personalities," *Nature* 447 (2007): 581–584.

14 The literature on this is enormous. See for example M. Nielsen et al., "A longitudinal investigation of self-other discrimination and the emergence of mirror self-recognition," *Infant Behavior and Development* 26 (2003): 213–226.

15 T. Suddendorf and E. Collier-Baker, "The evolution of visual self-recognition: Evidence of absence in lesser apes," *Proceedings of the Royal Society of London B* 276 (2009): 1671–1677; J. M. Plotnick et al., "Self-recognition in an Asian elephant," *Proceedings of the National Academy of*

Sciences of the USA 103 (2006): 17053–17057; D. Reiss and L. Marino, "Mirror self-recognition in the bottlenose dolphin: A case of cognitive convergence," *Proceedings of the National Academy of Sciences of the USA* 98 (2001): 5937–5942; H. Prior et al., "Mirror-induced behavior in the magpie (*Pica pica*): Evidence of self-recognition," *PLOS Biology* 6 (2008): e202, doi:10.1371/journal.pbio.0060202.

16 See for example J. L. Brown and A. M. Eklund, "Kin recognition and the major histocompatibility complex: An integrative review," *American Naturalist* 143 (1994): 435–461.

AFTERWORD

1 I'm thinking of *Independence Day*, a popcorn movie, which I reviewed somewhat witheringly in *Nature* 386 (1996): 681.

2 The first person to show me this paragraph taken out of context by creationists will be featured on my blog.

3 See for example R. L. Mancinelli et al., "Biopan-survival I: Exposure of osmophiles *Synechococcus* sp. (Nageli) and *Haloarcula* sp. to the space environment," *Advances in Space Research* 22 (1998): 327–334.

Index